Optically Pumped Magnetometer and Its Application

Optically Pumped Magnetometer and Its Application

Guest Editors

Jixi Lu

Yao Chen

MDPI

Basel • Beijing • Wuhan • Barcelona • Belgrade • Novi Sad • Cluj • Manchester

Guest Editors

Jixi Lu
School of Instrumentation and Optoelectronic Engineering
Beihang University
Beijing
China

Yao Chen
School of Mechanical Engineering
Xi'an Jiaotong University
Xi'an
China

Editorial Office
MDPI AG
Grosspeteranlage 5
4052 Basel, Switzerland

This is a reprint of the Special Issue, published open access by the journal *Photonics* (ISSN 2304-6732), freely accessible at: https://www.mdpi.com/journal/photonics/special_issues/401S361028.

For citation purposes, cite each article independently as indicated on the article page online and as indicated below:

Lastname, A.A.; Lastname, B.B. Article Title. *Journal Name* **Year**, *Volume Number*, Page Range.

ISBN 978-3-7258-4693-1 (Hbk)
ISBN 978-3-7258-4694-8 (PDF)
https://doi.org/10.3390/books978-3-7258-4694-8

Cover image courtesy of Jixi Lu

Contents

About the Editors

Jixi Lu

Jixi Lu is currently an Associate Professor at the School of Instrumentation and Optoelectronic Engineering, Beihang University, China. His primary research interests include atomic magnetometers, high-precision magnetic field control, and low-noise magnetic shields. His group has developed various kinds of miniature atomic magnetometers, which have achieved a sensitivity of $fT/Hz^{1/2}$ level. They have carried out applications such as extremely weak magnetic detection, metrology, and imaging. His research is funded by over 10 projects, including the Innovation Program for Quantum Science and Technology and the National Natural Science Foundation of China. He has published more than 50 peer-reviewed papers, 30 authorized patents, and has been awarded a first prize at the Ministry level.

Yao Chen

Yao Chen is currently an Associate Professor at the School of Instrument Science and Technology, Xi'an Jiaotong University, China. He got his PhD degree from Beihang University and had long study experience with an atomic co-magnetometer. He has postdoc experience in Professor Gerald Gabrielle's group at Harvard University, which focuses on the precision measurement of antiproton magnetic moment. Yao Chen's group's primary research interests include chip-scale quantum sensors, which may be applied in several areas, such as ultra-sensitive brain magnetic field sensing, ultra-low field NMR for molecular analysis, and low-drift quantum gyroscope. He is also studying trapped ions, which could form a quantum oscillator or rotor.

Preface

In recent years, quantum precision measurement technology has developed rapidly and obtained important applications in many fields. They have demonstrated extremely high accuracy in the measurement of physical quantities, such as time, gravity, magnetic fields, and electric fields. The optically pumped magnetometer (OPM) is a common type of quantum magnetometer, which relies on the measurement of Larmor precession of an atomic spin ensemble in the magnetic field. They need specific frequency light interacting with atomic ensembles to complete the pumping and detection process. Classical OPMs typically work under the Earth's magnetic field or a limited field, achieving scalar measurement of the magnetic field by measuring the Larmor precession frequency. About 20 years ago, researchers at Princeton University discovered that if OPMs were operating in near near-zero field, they could realize the spin-exchange relaxation-free (SERF) regime. Under this state, a substantial increase in sensitivity can be achieved. In addition, other types of OPMs have been developed, including nonlinear magneto-optical rotation (NMOR) magnetometers, coherent population trapping (CPT) magnetometers, radio-frequency magnetometers, etc. Each of them has unique characteristics and can meet the requirements of different application scenarios. At present, OPMs have been widely used in magnetoencephalography, magnetocardiography, geomagnetic detection, and inertia measurement (comagnetometer). There are a large number of researchers engaged in the study of relevant mechanisms, devices, technologies, and applications. More importantly, supported by MEMS technology and micro/nano optics, OPMs have the potential to move toward chip-scale sensors. This reprint originates from the special issue "Optically Pumped Magnetometer and Its Application" from the *Photonics* journal. Guest editors of this special issue have long been engaged in research on OPMs and have continuously paid close attention to the latest advancements in this field. Therefore, they invited relevant scholars to report their research in this Special Issue. They provide valuable references for improving the performance of atomic magnetometers or for meeting special application requirements. We hope the reprint can promote innovative research for researchers in the related field.

Jixi Lu and Yao Chen
Guest Editors

photonics

MDPI

Article

Theoretical Study on Performing Movement-Related MEG with ^{83}Kr-Based Atomic Comagnetometer

Yao Chen *, Ruyang Guo, Jiyang Wang, Mingzhi Yu, Man Zhao and Libo Zhao *

School of Instrument Science and Technology, Xi'an Jiaotong University, Xi'an 710049, China; 3123301344@stu.xjtu.edu.cn (R.G.); jiyangwang@stu.xjtu.edu.cn (J.W.); yumingzhi@stu.xjtu.edu.cn (M.Y.); zm1994@stu.xjtu.edu.cn (M.Z.)

* Correspondence: yaochen@xjtu.edu.cn (Y.C.); libozhao@xjtu.edu.cn (L.Z.)

Abstract: A K–Rb–^{83}Kr-based atomic comagnetometer for performing movement-related Magnetoencephalography (MEG) is theoretically studied in this paper. Parameters such as the spin-exchange rates, the spin-dephasing rates and the polarization of the nuclear spins are studied to configure the comagnetometer. The results show that the nuclear spin can generate a magnetic field of around 700 nT, at which the nuclear spin can compensate for a wide range of magnetic fields. In this paper, we also show the fabrication process for hybrid optical-pumping vapor cells, whereby alkali metals are mixed in a glove box that is then connected to the alkali vapor-cell fabrication system.

Keywords: atomic comagnetometer; spin-exchange optical pumping; optically pumped magnetometer; MEG; atomic spin gyroscope

1. Introduction

Magnetoencephalography (MEG) recordings have found various applications, including source localization of epilepsy-related seizures to aid in planning epilepsy surgery [1,2]; the study of the brain's response to specific external stimuli, which helps map the motor system [3]; and sensory areas [4], such as language, vision, etc. MEG signal recordings can also be used for the diagnosis of early-stage Alzheimer's disease [5] and the study of cerebral networks in tremor syndromes [6]. The typically used traditional MEG measurement equipment includes SQUIDs (super-conducting quantum interference devices), in which a very large helmet with liquid helium is used, which limits the application of magnetometers to the motor system, therefore, it is not feasible to perform MEG while a person freely moves their head [7,8].

With the development of the optically pumped magnetometer (OPM), which can work at room temperature and reach sensitivity equal to that of SQUID magnetometers [9,10], future MEG equipment might be wearable [11–13]. A wearable OPM system and EEG (electroencephalography) have been combined for the study of brain function [14]. This could reinforce more applications of MEG equipment, such as motor system-related space navigation of the brain and MEG for visual–motor integration.

Even though OPMs are wearable and movable, an important technique-related problem should be considered. Different from EEG, in which there is no large background electric field in the environment, MEG recording could be affected by the geomagnetic field of the earth. We know that the brain magnetic field is quite weak compared with the geomagnetic field. Thus, magnetic-field shield rooms (MSRs) are typically used for suppression of the environmental magnetic field. Moreover, the penetration of the magnetic field into MSRs can still affect MEG recording. In order to perform movement-related MEG, a large bi-planar coil was used for movement-related background-magnetic-field compensation [15]. During MEG recording, a person can only move in a small area, or only very small movements can be performed, such as drinking water [8]. In order to enable ambulatory movements in wearable MEG, matrix coil active magnetic shielding was used [16]. Some other methods based on magnetic-field compensation have also been studied [17–19]. The magnetic-field differential method can also be used for background-magnetic-field suppression [20].

Even though many methods have been developed for movement-related MEG measurement, most of these methods are based on magnetic-field compensation in the space of the movement. Large compensation coils are required, which can make the system very complicated. Moreover, this method cannot compensate for the magnetic field of each sensor head, and some of the sensors could be over- or under-compensated.

As we understand, the crucial aspect of performing movement-related MEG lies in compensating for fluctuations in the magnetic field. The method based on hyperpolarized nuclear spins can also be used [7]. The nuclear spin method could compensate for the magnetic field in situ as well as automatically compensate for the fluctuations in the magnetic field in each of the OPMs. Complicated compensation coils are not needed, resulting in a significant reduction in the system's volume.

Hyperpolarized nuclear spins include those of ^{3}He,^{21}Ne,^{129}Xe and ^{131}Xe. The first demonstration of MEG measurement during movement was performed with ^{21}Ne [7]. There is a small spin-exchange rate between ^{21}Ne and alkali atoms such as K or Rb and the ^{21}Ne nuclear spin features a very large quadrupolar relaxation rate [21]. In addition, ^{21}Ne nuclear spins could not be easily polarized and hybrid pumping technique is required [22]. Moreover, the vapor cell temperature could reach 473 K for the polarization. The typically used boro-silicate glass could not withstand such high temperatures due to glass corrosion. The ^{21}Ne nuclear spins are widely used for rotation sensing or the studying of fundamental physics [23–27]. However, due to the very low natural abundance of isotope-enriched ^{21}Ne, it is very expensive to buy this gas. The ^{3}He nuclear spins feature a very small spin-exchange optical-pumping rate with alkali metals; thus, it is very hard to polarize the nuclear spins of ^{3}He. A hybrid pumping technique may be used. However, there is no report about the hybrid pumping of ^{3}He for applications with atomic comagnetometers. Moreover, ^{3}He nuclear spins are quite sensitive to magnetic-field gradients and can be easily depolarized in a holding magnetic field. Even the stem of a vapor cell could affect the spin polarization of nuclear spins [28–30]. It is worth noting that ^{129}Xe and ^{131}Xe can also be used. However, due to the large collision relaxation at the container walls, it is hard to polarize the nuclear spins of these elements [31,32]. For ^{83}Kr nuclear spins whose atom mass is much smaller than that of Xe atoms, we can estimate smaller wall relaxation as well as weaker interaction strength for these alkali atoms. Moreover, ^{83}Kr is easier to polarize than that of the ^{3}He and ^{21}Ne. In addition, ^{83}Kr is a good candidate for movement-related MEG measurement. In this paper, we will focus on the study of ^{83}Kr nuclear spins for background-magnetic-field fluctuation compensation in movement-related MEG measurement. We will first study the configuration of the ^{83}Kr-based comagnetometer. The relaxation of ^{83}Kr should be studied first since the relaxation could determine the polarization of the nuclear spins. Then the polarization of the nuclear spins is calculated. In order to acquire the sensitivity of the nuclear spin detection, we studied the Rb relaxation. The sensitivity of the comagnetometer will be calculated in the "Configuration of the Comagnetometer" section. In the Section 3, we studied the hybrid pumping cell fabrication. As we know, before we study the atomic comagnetometer, we need to first fabricate an alkali vapor cell. Hybrid pumping cells are needed. Thus, we have shown the method for hybrid pumping cell fabrication. The Sections 4 and 5 of the paper are the "Discussion" and "Conclusions".

2. Configuration of the Comagnetometer

2.1. Relaxation of ^{83}Kr

The nuclear spin-based field compensation method for movement-related MEG is heavily related to nuclear spin polarization, which is closely related to the spin relaxation of nuclear spins. The ^{83}Kr atom has a nuclear spin of $I = 9/2$; thus, it has a quadrupolar moment. We can say that quadrupolar relaxation is the main relaxation process of these nuclear spins. Moreover, container-wall-related quadrupolar relaxation of nuclear spins and quadrupolar relaxation in

^{83}Kr–^{83}Kr collisions are the main quadrupolar relaxation processes. There are several classical papers which illustrated the relaxation of ^{83}Kr in detail; see [32–34].

Quadrupolar relaxation in ^{83}Kr–^{83}Kr collisions is related to the number density of ^{83}Kr atoms. The higher the number density of the nuclear spins, the faster the relaxation that is caused by quadrupolar interaction. It is reported that the relationship between quadrupolar relaxation time T_1 and number density of the nuclear spins ρ, in Amagat (1 Amagat is $2.69 \times 10^{19}/\mathrm{cm}^3$), is $1/T_1 = \rho(2.13 \pm 0.05) \times 10^{-3}\mathrm{s}^{-1}$ [34]. Under typical conditions of 50 Torr gas, the relaxation rate of gas-phase ^{83}Kr collisions is $1.4 \times 10^{-4}\mathrm{s}^{-1}$.

Quadrupolar relaxation at the container wall and spin-exchange relaxation of alkali atoms are the main relaxation processes and are around an order of magnitude higher than the ^{83}Kr–^{83}Kr collision process. Since ^{83}Kr has a quadrupolar moment, the nuclear spins can be relaxed by having them collide with the container wall of the vapor cell, where the impurities in the wall could produce electric-field gradients (EFGs). The spin-exchange optical-pumping process of alkali atoms can also cause the decay of nuclear spin polarization. The spin momentum can transfer between alkali atoms and ^{83}Kr atoms. Note that the spin-exchange process can also be the pumping process, which can make the nuclear spins become polarized. We define the total relaxation rate of ^{83}Kr nuclear spins [32] as

$$(\frac{1}{T_2})_{m,n} = f_{m,n} < \omega_Q^2 > J(0) + \sigma_{ex}\bar{v}n_{Rb}, \tag{1}$$

where m and n represent the energy level, and in our study, we typically consider the m = 1/2 and n = −1/2 sublevels. For ^{83}Kr nuclear spins, we calculate $f_{m,n} = 5/4$ using Equation (118) in reference [35].

$$< \omega_Q^2 >= (\frac{3e^2Q(1-\gamma_\infty)}{2\hbar I(2I-1)})^2 < q^2 >, \tag{2}$$

which includes the mean square of the local quadrupole coupling constant ($< e^2qQ$ $(1-\gamma_\infty)h^{-1} >^2$) as a measure of the mean interaction strength at the surface-wall site. In the reference, the coupling constant is measured to be 5.6 MHz for ^{83}Kr atoms on typical Duran glass [32]. I is the nuclear spin quantum number of ^{83}Kr. $\hbar$ is Planck's constant. Note that q is equal to $\sqrt{< q^2 >}$, which is defined to be the root mean square of the electric-field gradient. With these parameters, we can calculate $< \omega_Q^2 >$, which is approximately not temperature-dependent. $J(0)$ is defined to be the spectral density function of the normalized correlation function at zero frequency [36]. It is related to the stochastic sequence of wall collisions and thermally activated surface diffusion. Under our experimental conditions, it is believed that diffusion along the surface is strongly hindered and that quadruple relaxation is governed by the adsorption and desorption processes.

$$J(0) \approx \frac{\tau_s^2}{\tau_v} \propto exp(2E_A/k_BT) \tag{3}$$

In the above equation, τ_v equals $\bar{v}S/4V$, which is the average time it takes the nuclear spin to move to the wall. $\bar{v} = \sqrt{8k_BT/\pi M}$ is the average velocity of the nuclear spins. M is the atom mass of the noble gas atom. V is the volume of the vapor cell, and S is the overall surface of the vapor cell. k_B is the Boltzmann constant. E_A is the activation energy of desorption. We cannot directly obtain τ_s, which is defined as the adsorption time of the nuclear spins on the surface wall. However, we can use the squared phase angle, $< \theta^2 >$, as in the reference [32], to acquire τ_s. Using the relation $\tau_s^2 = (\tau_0 exp(2E_A/k_BT))^2 =< \theta^2 > / < \omega_Q^2 >$, we can calculate τ_s with the measured $< \theta^2 >$ at 373 K as $4.9 \times 10^{-8} rad^2$. Note that τ_v is related to $\bar{v}S/4V$. The cell shape and dimensions could determine the relaxation rate. In our simulation, we suppose that $4V/S$ is equal to 10 mm for approximation. This value is typical for vapor cells utilized in the hyper polarization experiments. It is also approximately equal to values in reference [32].

The total relaxation of ^{83}Kr also includes the spin-exchange interaction with Rb atom spins, which will be considered in our study. The spin-exchange interaction cross-section (σ_{ex}), the velocity ($\bar{v}$) and the number density of Rb (n_{Rb}) determine the relaxation rate. Based on the spin-exchange term $\sigma_{ex}\bar{v}^2 = 1.9 \times 10^{-12}\mathrm{cm}^4\mathrm{s}^{-2}$ measured in the reference [32], we can obtain a numerical simulation of the total relaxation rate, as shown in Figure 1. The smallest relaxation rate happen at the temperature of 382 K and the relaxation rate is approximately 0.0021 s^{-1}. The relaxation time is around 500 s. This relaxation time is coincidental with the relaxation times acquired in these references [32,33]. Note that here we only consider the collision relaxation with Rb atoms and without considering the collision with the K atom spins. We have mentioned that K-Rb hybrid pumping would be utilized in the comagnetometer.

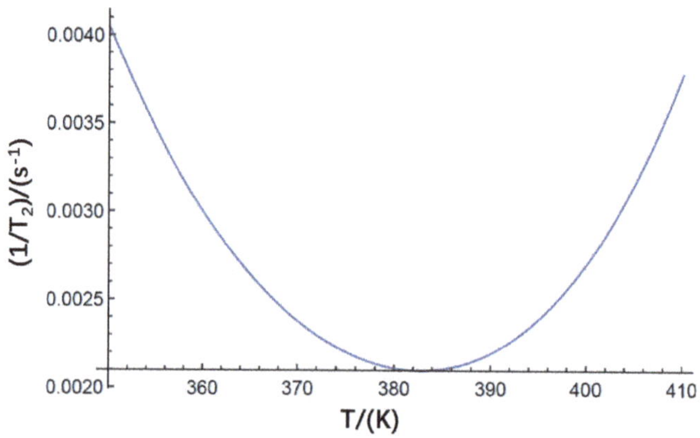

Figure 1. Numerical modeling of the dependent relaxation rate of nuclear spins on the temperature of the vapor cell. At low temperatures, ^{83}Kr is adsorbed on the surface walls for a longer time, and fluctuation EFGs cause the relaxation of the nuclear spin. As the temperature increases, relaxation among collisions with Rb atoms is larger, because the number density of Rb is larger, and more alkali atoms depolarize the nuclear spins. The spin-exchange relaxation is dominant at high temperatures, since the atoms stay on the surface for a short time and the quadrupolar relaxation can be neglected.

In a hybrid optical pumping experiment, a mixture of the metals is typically used. According to the Raoult's law, the partial pressure of the saturated vapor of a metal above a liquid phase is determined by the composition of the mixture or the mole fraction ratio of a metal in the mixture. The partial pressure of a metal can vary from 0 to the value of the pure vapor pressure. As we know, the partial pressure of the metal could determine the density of the metal in the vapor cell. Thus, the density ratio of Rb to K could be determined. In the hybrid pumping experiment for the atomic comagnetoemter in which very large nuclear spin polarization is required, the density ratio of Rb to K is typically very large which means that most of the alkali atoms are Rb. The density ratio of Rb to K is more than 200 in these experiments [37].

In the hybrid pumping experiment, K is directly pumped by the laser and Rb is pumped by spin exchange interaction with K electron spins. As have mentioned that the K density ratio is quite smaller than that of the Rb metal in our vapor cell design, the K medium could be designed to be optically thin (the optical depth is typically from 1 to 2, the temperature of the vapor cell could be changed to reach the required optical depth). Due to the large density ratio of Rb to K, the optical depth of Rb is larger than that of K. This means that Rb could not be directly pumped efficiently by a laser. The polarization gradient of Rb in the laser pass is quite large. However, due to the large collision cross section of the spin exchange interaction between K and Rb, the spin transfer rate from K to Rb is quite large. Rb could be uniformly polarized under hybrid pumping conditions. The polarization gradient could greatly affect the spin polarization of the nuclear spin. A large polarization gradient could lead to the failure of an efficient nuclear spin hyper polarization. Note that even though we could allow the laser power to be as large as possible to reach a polarization of 1, we need to keep the alkali metal electron spin polarization at 0.5 to obtain the best sensitivity of the comagnetometer. We could see that in the hybrid pumping experiment, the Rb atomic spins are mainly used for spin exchange optical pumping of the nuclear spins. Due to the large density of Rb, some of the nuclear spins, such as ^{3}He and ^{21}Ne, which are very hard to polarize, could be efficiently polarized. It is reasonable to only consider the Rb-related collision relaxation in Equation (1).

2.2. Polarization of ^{83}Kr

The polarization of the nuclear spin is very important, because it can determine the magnetic field experienced by the electron spins. With the relaxation and the spin-exchange optical-pumping rate of alkali atoms and nuclear spins, we can determine the polarization of the nuclear spins. In Equation (1), the term $\sigma_{ex}\bar{v}n_{Rb}$ can also be defined to be the spin-exchange optical-pumping rate of Rb and nuclear spins. The polarization of the nuclear spins can be defined as

$$P^n = P^e \frac{\sigma_{ex}\bar{v}n_{Rb}}{f_{m,n} < \omega_Q^2 > J(0) + \sigma_{ex}\bar{v}n_{Rb}}, \tag{4}$$

where P^e is the polarization of the electron spin. We can see that the highest polarization of the nuclear spin can reach that of electron spins. It is known that the polarization of the electron spin can be 0.5, at which the magnetometer has the highest sensitivity. Thus, we let P^e be 0.5. We performed a numerical simulation of the polarization of nuclear spins with the change in temperature. Figure 2a shows the results.

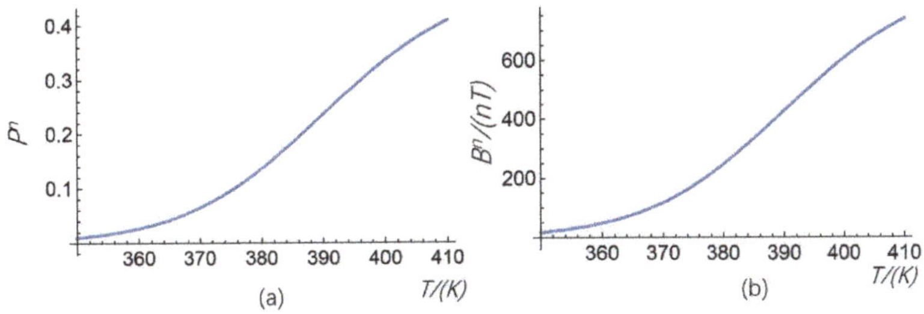

Figure 2. (**a**) Theoretical estimation of the relationship between temperature and nuclear spin polarization (P^n). As the temperature of the vapor-cell changes, we can estimate a rise in the spin-exchange optical pumping of the nuclear spins as well as a decrease in quadrupolar relaxation. Thus, the polarization of nuclear spins increases with the rise in temperature. (**b**) Theoretical estimation of the relationship between the temperature and the Fermi contact interaction enhanced magnetic field produced by the nuclear spin (B^n). As the magnetic field produced by the nuclear spin is directly related to nuclear spin polarization, we can estimate a similar change in B^n with temperature.

As shown in Figure 1, the lowest relaxation falls at 382 K. We need to choose the appropriate temperature to minimize the relaxation rate as well as achieve very high nuclear spin polarization. Even though the lowest relaxation rate occurs at 382 K, however, as shown in Figure 2a, the polarization at 382 K is around 0.15—far less than the highest polarization, which is around 0.4. Thus, it is better for us to choose a working temperature at which the polarization of the nuclear spin is relatively higher as well as the boro-silicate glass could withstand such temperatures without corrosion of the glass. It is better for us to choose the temperature around 420 K.

In order to compensate for the fluctuation in the background magnetic field experienced by electron spins used in brain magnetic-field sensing, it is necessary to calculate the magnetic field generated by the nuclear spins (B^n). We can estimate that the larger the nuclear spin magnetic field is, the higher the compensation ability of the nuclear spins is. The magnetic field produced by the nuclear spins is defined as

$$B^n = \frac{8}{3}\pi k_0 \mu_{Kr} n_{Kr} P^n, \tag{5}$$

where k_0 is the Fermi contact interaction enhancement factor, equal to 270 for the $Rb -^{83} Kr$ pair [38,39]. μ_{Kr} is the magnetic moment of the ^{83}Kr nuclear spin. n_{Kr} is the number density of the nuclear spins. We supposed that the partial pressure of ^{83}Kr gas inside the vapor cell was 50 Torr. We performed a simulation of the nuclear spin magnetic field with the change in temperature. Figure 2b shows the results.

The polarized nuclear spin could produce magnetic field B^n. This field could compensate the background disturbing field B_y or B_x. As shown in Figure 3, the pump laser polarized the electron spins in the z direction and the electron spins could produce a similar magnetic field B^e as that of the nuclear spin B^n. We can add a holding magnetic field which is constant in the z direction and the direction of this field is opposite to B^n and B^e. The strength of this field is equal to $B^n + B^e$. Under this condition, the nuclear spin could automatically compensate the disturbing field in the x or y direction. If there is no disturbing field, both B^n and B^e will settle in the z direction. Due to the Fermi contact interaction, the nuclear spins could experience B^e. The total magnetic field experienced by the nuclear spin would be $B_c\vec{z} + \vec{B}^e$. The net magnetic field experienced by the nuclear spin could be the same as the nuclear spin magnetic field B^n and the only difference is that the direction is opposite. If there is a y disturbing magnetic field B_y input, a total magnetic field $\vec{B}_{tot}$ will be produced and the nuclear spin will also precess to a new direction which is approximately opposite to the total magnetic field direction. The projection of $\vec{B}^n$ in the y direction would be B^n_y. We can calculate that B^n_y is equal to B_y. This means that the nuclear spins have compensated the disturbing field B_y. The electron spins will stay in the z direction and there will be no signal output in the comagnetometer. Note that the electron spins are used to detect the magnetic field produced by the brain or the background magnetic field. However, if there is a brain's magnetic field input whose frequency is very large compared to the comagnetometer's compensation bandwidth, the comagnetometer may not be fast enough to compensate for it. Thus, the comagnetometer could be used for the measurement of the brain's magnetic field as well as being insensitive to the disturbing low-frequency background magnetic field.

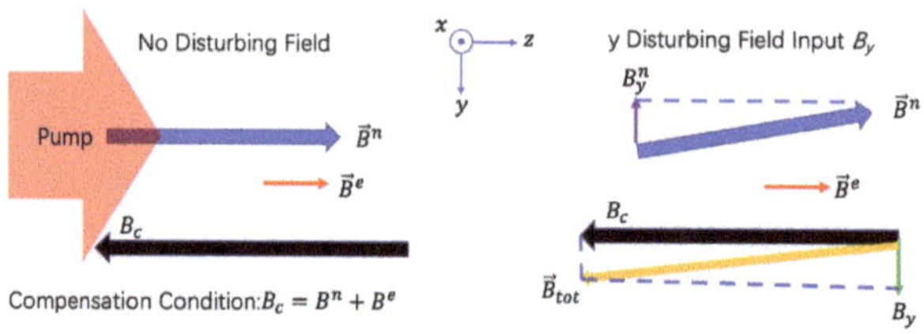

Figure 3. The principle of disturbing field compensation by the polarized nuclear spins.

The disturbing magnetic field suppression should also be studied in this paper. The nuclear spin could compensate the disturbing magnetic field. To illustrate this process, we assume that there is sinusoidal disturbing magnetic field input in the comagnetometer $B_y cos(2\pi f t)$ in which f is the frequency of the disturbing magnetic field. The signal output of the comagnetometer S_y is equal to $Ksf_y B_y cos(2\pi f t + \Phi_y)$. Here, Φ_y is the relative phase between the input and the output signal. K is a factor that links the input magnetic field and the final comagnetometer output voltage signal and it is constant. If there is no suppression process, which means no nuclear spins are used, the signal output could be

$KB_y cos(2\pi ft + \Phi_y)$. The factor sf_y is defined to be the suppression factor of the B_y field which (see [25,40]) is:

$$sf_y = \frac{\omega^2}{[\omega_n^2 + \omega^2\omega_e^2/(R_{tot}/\gamma^e)^2]}, \tag{6}$$

where ω is the angular frequency of the input magnetic field and it is equal to $2\pi f$, ω_n is equal to $\gamma^n B^n$ and γ^n is the gyro-magnetic ratio of the nuclear spin, ω_e is equal to $\gamma^e B^e/Q(P^e)$, γ^e and $Q(P^e)$ are the gyro-magnetic ratio of the electron spin and the slow down factor [41], respectively, and R_{tot} is the total relaxation of the electron spins which are shown in the following section. Similarly, we can define the x direction magnetic field suppression factor:

$$sf_x = \frac{\omega}{[\omega_n^2 + \omega^2\omega_e^2/(R_{tot}/\gamma^e)^2]^{1/2}}. \tag{7}$$

To further show the compensation process, we show a picture in Figure 4. Here, we chose the typical parameters whereby the temperature of the vapor cell is 420 K. Other parameters such as the nuclear spin polarization and the nuclear spin magnetic field B^n could be determined. We chose the electron spin polarization to be typically 0.5 under which the comagnetometer could utilize the best sensitivity. The magnetic field produced by the electron spins could be:

$$B^e = \frac{8}{3}\pi k_0 \mu_B n_{Rb} P^e, \tag{8}$$

in which μ_B is the Bohr Magneton and n_{Rb} is the number density of Rb. The relation between the frequency of the disturbing magnetic field and the suppression factors are shown. As we can see in the frequency range of 1 to 100 Hz, the responses of the comagnetometer to magnetic field is flat and there is about three times the suppression of the magnetic field in the x direction. Moreover, as the frequency changes decrease in size, the suppression factor decreases and the magnetic field is suppressed more efficiently. For the y direction, the magnetic field suppression is larger.

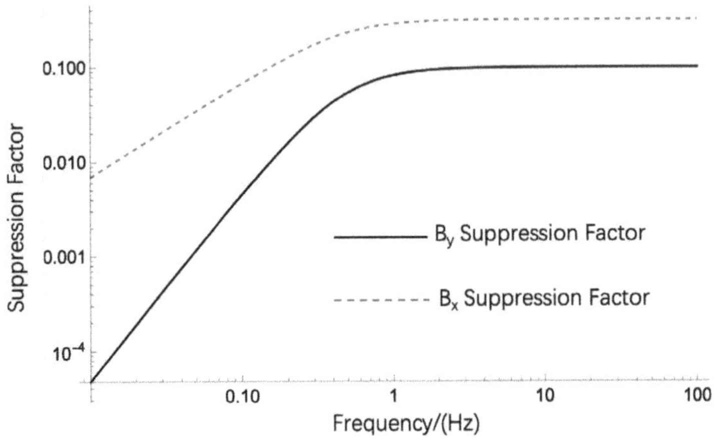

Figure 4. Theoretical simulation of the relationship between magnetic field suppression factors of the comagnetometer and the input magnetic disturbing magnetic fields' frequencies.

2.3. Relaxation of Rb and Fundamental Sensitivity Estimation

The relaxation of Rb needs to be considered, because it is related to the spin projection noise of comagnetometers. Since Rb is used for the detection of the brain's magnetic field as well as the nuclear spins' compensation magnetic field, the sensitivity of a magnetometer is largely determined by the relaxation of Rb. The relaxation of Rb is caused by several collision processes, such as collisions with itself, collisions with N_2 quenching buffer gas and collisions with ^{83}Kr nuclear spins. These collision relaxation rates are closely related to the collision cross-sections and number densities of gases. We performed a simulation of the relaxation rates, and Figure 5 shows the results. We changed the temperature of the vapor cell; thus, both Rb–Rb spin destruction relaxation and Rb–^{83}Kr spin-exchange relaxation changed with the temperature. The spin destruction rate of Rb–N_2 resulted to be weakly temperature-dependent. We also show the relationship between the temperature and the Rb–^{83}Kr spin-exchange relaxation rates.

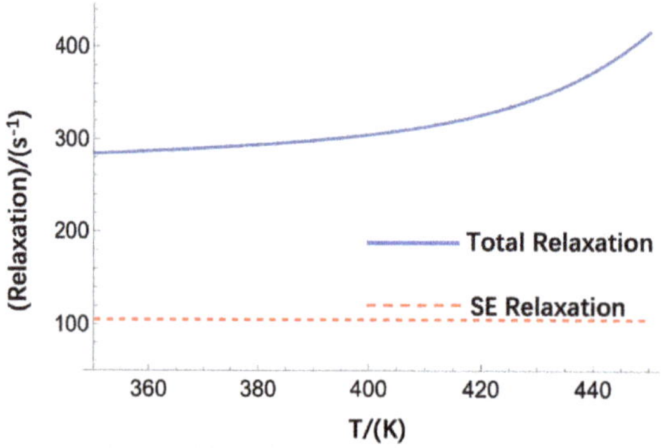

Figure 5. Theoretical simulation of the relationship between temperature and the relaxation of Rb, which is used to polarize the nuclear spins. 'SE Relaxation' stands for 'Spin Exchange Relaxation'. Total relaxation includes spin destruction relaxation between Rb and N_2, spin destruction relaxation between Rb and Rb, and the spin transfer rate from Rb to ^{83}Kr. SE relaxation only includes the transfer rate of the Rb electron spin to the nuclear spins.

After we have obtained the total relaxation of Rb electron spins, we can calculate the fundamental sensitivity of a comagnetometer, which is limited by the spin projection noise. The fundamental sensitivity of the comagnetometer is limited by the total relaxation of the atomic spins, the number density of the spins and the volume of the vapor cell. The fundamental sensitivity of a comagnetometer is defined as δB:

$$\delta B = \frac{1}{\gamma^e}\sqrt{\frac{Q(P^e)2R_{tot}}{n_{Rb}V}}, \tag{9}$$

where γ^e is the gyro-magnetic ratio of the electron spin. $Q(P^e)$ is the slow-down factor of Rb electron spins [9]. R_{tot} is the total relaxation of Rb spins, which is calculated in the above subsection. Note that there is a factor of 2 here. This factor here means that the optical pumping was also considered. As the optical pumping rate is equal to the total relaxation rate of Rb, the comagnetometer utilizes the best signal-to-noise ratio or sensitivity. Thus, a factor of 2 here considered both optical pumping as well as relaxation's contribution to the spin projection noise. We know that Rb was not directly pumped by the laser. However, we can calculate an equivalent pumping rate which is approximately equal to the optical pumping rate of K divided by the density ratio of Rb to K [42]. n_{Rb} is the number density of Rb, which is related to the temperature of the vapor cell. V is the interaction volume of Rb atoms and laser light. With the equation, we can calculate the fundamental sensitivity of a comagnetometer with changes in temperature and interaction volume. Figure 6 shows the results. We can see that even in a very small volume (0.002 cm^3), we can obtain more than 10 fT estimated sensitivity if the temperature of the vapor cell is high enough. This means that we can fabricate very small sensor heads for MEG detection. The space resolution is quite high if we use relatively high temperatures for the vapor cell.

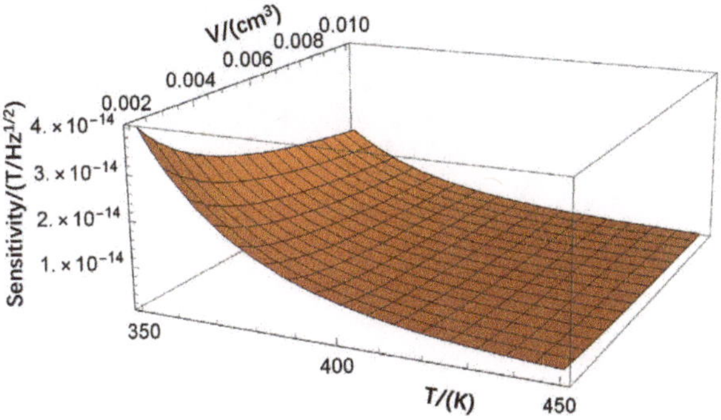

Figure 6. Theoretical simulation of the dependent fundamental sensitivity of the comagnetometer on the temperature and the active volume of the vapor cell. The temperature of the vapor cell was changed; thus, the number density of Rb atoms changed. As the temperature increased, we estimated higher sensitivity. We also changed the active volume of the vapor cell in which laser light interacts with Rb atom spins; then, the fundamental estimated sensitivity of the comagnetometer was calculated. As the volume increased, the fundamental sensitivity increased.

Note that the suppression factor could also contribute to the sensor's sensitivity. There is approximately a suppression factor of 3 for the x direction magnetic field in the frequency range of 1–100 Hz. Thus, we believe that the sensitivity of the comagnetometer for detection of the x magnetic field could also be suppressed by a factor of 3. The y direction suppression factor is much larger and thus we think that the comagnetometer is not sensitive to the y magnetic field and it could be treated as a single axis magnetometer. Even though there are some suppression factors for magnetic field detection, we can improve the fundamental sensitivity through increasing the interaction volume of the cell as well as the temperature of the vapor cell.

3. Cell Fabrication

3.1. K–Rb Mixture Preparation

In order to achieve very high nuclear spin polarization as well as reduce the spin-exchange relaxation of Rb and ^{83}Kr nuclear spins, hybrid optical pumping is a good solution. If we want to achieve very high nuclear spin polarization, we need to increase the vapor-cell temperature. However, a high-temperature vapor cell could lead to very large optical depth; thus, the laser light might not easily pass through the medium. The polarization of nuclear spins might also be decreased. Hybrid optical pumping is thus needed. The hybrid optical pumping is usually applied in the comagnetometer because its efficiency is higher than a normal optical pump. There are alkali atoms and noble gas atoms in the vapor cell and the electron spin of the alkali metal and the nuclear spin of the inert gas are coupled to achieve hybrid optical pumping. The process is as follows: the electron spin is first pumped by the laser and the nuclear spin is hyperpolarized by spin-exchange collision. Therefore, the polarizability is greatly improved as well as the sensitivity of the comagnetometer. With this technique, we can achieve very high nuclear spin polarization.

We had to prepare the alkali vapor cell before the experiment could be performed. The hybrid optical-pumping cell had to be filled with a mixture of alkali metals. Before the vapor cell was filled with K–Rb mixture using a torch flame on the fabrication apparatus, the K–Rb mixture with the intended Mole Fraction Ratio (MFR) had to be prepared in the glove box. The glove box is an inert gas protection box that can be vacuumed to the main box, so that the system is kept in a high purity inert gas protection environment.

When preparing the K–Rb mixture, the mass ratio of K to Rb had to be determined first. The mass ratio is defined as $M_r = m_K / (m_{Rb} + m_K)$, where m_K is the mass of K and m_{Rb} is the mass of Rb. The mass ratio determines the MFR and the density ratio of K to

Rb in a hybrid pumping cell (HPC) [43]. In a $K - Rb -^{87} Kr$ comagnetometer, the density ratio (DR) of K to Rb is controlled for improving the uniformity of spin polarization at the operation temperature. For example, we prepared groups of mixtures in density ratios of 1/212 and 1/109 at 463 K after the cell was filled with them. In order to reach the intended DR, Table 1 shows the masses of K and Rb that had to be mixed together in the glove box.

Table 1. The calculation of the masses of K and Rb that had to be mixed together to reach the intended D_r.

Intended D_r@463 K	M_r	m_K (mg)	m_{Rb} (g)	MFR of K
1/212	0.014	30	2.13	0.029
1/109	0.026	41	1.54	0.055

The calculation of the mass of K and Rb determines the amount of alkali metal that should be mixed together in the glove box. Here, before the mixing of alkali metals, the glassware used in this experiment was cleaned before being put into the glove box to protect the alkali metals against impurities. All of the glassware was immersed into piranha solution and cleaned with an ultrasonic cleaner for 5 minutes. Several rounds of rinsing with deionized water were employed to rinse away the residual piranha solution. After cleaning it, we quickly put the glassware into a drying box and left it there for several hours to remove any water on the glassware. When the glassware was put into the transfer chamber of the glove box, the gases in the chamber were pumped away to further dry the glassware.

The mixing of K–Rb metals was performed after a low concentration of oxygen was achieved in the glove box. Commercial Rb and K break-seal ampoules were employed. A homemade glass spoon was used to scoop Rb and K metals out of the ampoules. With a glass knife, we cut the metals into small pieces in a Petri dish. A piece of weighting paper was put on the analytical balance to weigh the pieces of alkali metals. According the calculation results in Table 1, 30 mg of K and 2.13 g of Rb were added to a homemade break-seal ampoule to obtain 1/212 D_r at 463 K in a hybrid pumping cell. A Teflon valve and an O-ring sealed the alkali metal mixture in the ampoule to protect the mixture against air when the ampoule was taken out of the glove box. To seal the mixture in the glass ampoule and protect the mixture against air, a vacuum pump was used to pump out the gases in the ampoule. Methane flame was used to seal the glass at the neck of the ampoule during pumping. The mixture was finally sealed in the glass ampoule. Figure 7 shows this process. For HPCs with small MFRs, we used a glass knife to cut the alkali metal into small pieces of approximately 30 mg. Moreover, we used a valve-and-pumping technique to protect the mixture against oxygen and water after the ampoules were moved out of the glass box.

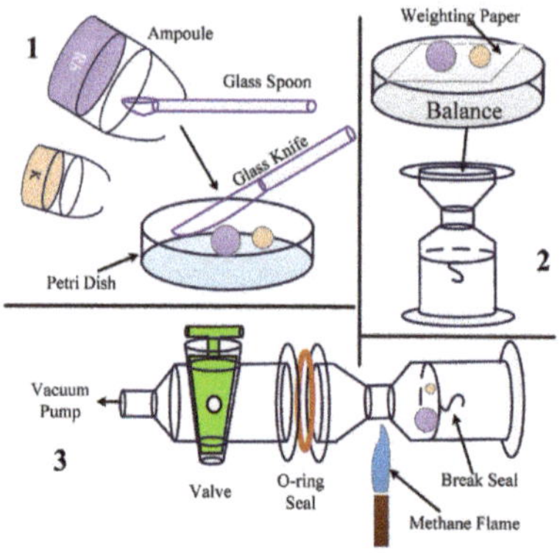

Figure 7. The process of experimentally preparing the K–Rb mixture in a glove box: 1. The cutting of the alkali metals. 2. The weighting of the alkali metals. 3. The sealing of the mixed alkali metals.

3.2. Alkali Chasing and Gas Filling

The mixture break-seal ampoule was attached to the alkali vapor-cell fabrication apparatus, and the cell fabrication process began when the ampoule was ready. Figure 8 shows the method of filling the vapor cells with the K–Rb mixture. The K–Rb mixture ampoule was connected to the alkali fabrication apparatus with a glass tube. An iron hammer, which was connected with the ampoule to break the seal, was sealed in the vertical glass tube. The vapor cell was connected to the glass tube, and the glass tube was connected to the vacuum pumping system and the gas line system. The vacuum pumping system pumped the whole system, and the gas line system supplied the vapor cell with gases such as ^{21}Ne, ^{131}Xe, ^{129}Xe, ^{83}Kr, N_2. The gas line system and the vacuum pumping system were similar to the system described in reference [44]. All the glassware was cleaned with piranha solution and deionized water, like the glassware used to make the mixture ampoule. To further clean the surface of the glassware, after it was connected to the fabrication system, an oven was used to heat the glassware to about 473 K when the pumping system was on. After 2 to 3 days of heating and pumping, the vacuum pressure reached the 10^{-5} Pa level. The vapor cells were filled with K–Rb mixture when the vacuum reached the 10^{-5} Pa level. Then, the K–Rb mixture was slowly chased into the retort using a flame for thoroughly mixing the K–Rb alloy and purifying the alloy. The ampoule fired off at the small neck of the glass tube after the mixture was chased into the retort. After that, the mixture was chased into the vapor cell, and the experimental results show that Rb moved faster than K due to the lower melting point of Rb. As a result, a special chasing method was developed to ensure that the MFR in the HPC was right. Most of the mixture was chased into the cell and then chased out of the cell to leave the amount of alloy that was needed. This process ensured that the MFR in the cell was close to the MFR in the ampoule. Figure 8 shows the schematic of the filling process.

After the vapor cells were filled with alkali metal mixtures, the vapor cells were filled with gases through the gas line. In some of the experiments, the vapor cell was filled with gases at about 2 Atm. For sealing the glass using a flame, the cell cannot be sealed until the inner pressure is lower than the pressure outside. Liquid nitrogen was used to fill the cell with gases at more than 1 Atm.

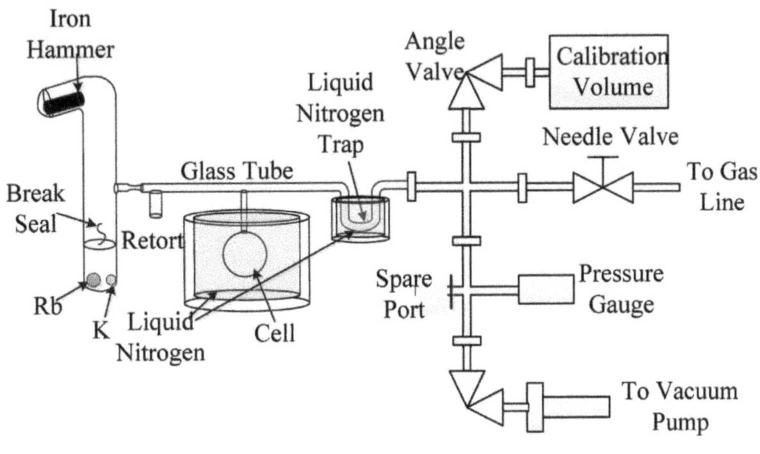

Figure 8. Schematic of filling the vapor cell with K–Rb mixture and gases experimentally.

Figure 8 shows the schematic of filling the glass cell with gas with a pressure more than 1 Atm. The liquid nitrogen trap was taken down, and the temperature of the trap rose to room temperature before gas filling. The needle valve allowed the gas to flow into the cell, and the pressure gauge measured the pressure in the cell. When the intended pressure (P_0) was reached, the gas line was shut down. A homemade Dewar was filled with liquid nitrogen, and the Dewar was lifted with a lifting platform to immerse the cell into liquid nitrogen. The pressure decreased and settled down to a lower pressure (P_1). A torch flame was used to fire off the stem of the cell, and the pressure rose slightly to P_2. This phenomenon was due to the increase in the temperature of the cell when the flame was used to fire off the cell from the glass tube. In order to know exactly the pressure of the cell after it was fired off from the glass tube and the temperature of the cell increased to room temperature, the ideal gas equation was used to calculate the pressure. The volume of the cell was measured to be V_c, and the volume of the glass tube and the gas line tube was measured to be V_g (the measurement method is described below). The pressure of the cell at room temperature should be

$$P_c = P_0 + (P_0 - P_2)\frac{V_g}{V_c} \tag{10}$$

Further, V_g was calculated with the ideal gas equation, and standard calibration volume was used for the calculation. We supposed that the standard calibration volume was V_{cal} and that the calibration volume was pumped to vacuum state. V_g was filled with gas, and the pressure was P_{g0}. After opening the angle valve of the calibration volume, the pressure changed to P_{g1}. The volume of the glass tube and the gas line was calculated to be

$$V_g = \frac{P_{g1} V_{cal}}{P_{g0} - P_{g1}} \tag{11}$$

After volume V_g was measured, cell pressure P_c was finally calculated with Equation (10). P_c is the pressure of the gases in the vapor cell at room temperature.

4. Discussion

This paper theoretically studies a ^{83}Kr-based atomic comagnetometer, which is a promising device for movement-related MEG recording. The main applications of moving MEG include the diagnosis of neuropsychiatric diseases such as epilepsy and Alzheimer's disease, the determination of brain function injury, and the localization of brain function before craniocerebral surgery. Moving MEG can also be used to explore brain functions, such as what people are thinking while drinking coffee or playing the guitar.

Due to the large Fermi contact interaction with Rb electron spins, the compensation field produced by the nuclear spin is quite large. A wide compensation range can be achieved, which is of utmost importance for effectively compensating for fluctuations in background magnetic fields in the future. Due to the small destruction collision relaxation rate of ^{83}Kr atoms, Rb magnetometers could achieve very high sensitivity. An experimental setup is under way to configure a K–Rb–^{83}Kr comagnetometer.

The temperature of the vapor cell designed in this study is 420 K and around 760Torr N_2 is filled in the vapor cell. The condition here is quite different from the typical ^{83}Kr polarization experiment in which the temperature of the vapor cell is around 373 K and only several tens of Torr N_2 is filled in the vapor cell. The polarization of ^{83}Kr is typically smaller than 10% [33,45]. With the high density of Rb, the polarization of ^{83}Kr could nearly reach that of the Rb polarization. The high pressure of N_2 in our design also eliminates the relaxation of Rb-^{83}Kr due to the formation of van de Waals molecules. The characteristic spin-relaxation rate due to the van der Waals molecule γ_M is only 63 sec^{-1} [33] which is much smaller than that of the Rb-Xe pair [46]. That is because Rb-Xe pair could form van der Waals molecules more easily. N_2 could also affect the formation of the molecule and we fill the high pressure of N_2 to stop the formation of the molecules in our design. According to this reference [46], the spin exchange rate coefficient due to van der Waals molecule κ_T is $(\gamma_M\zeta)/([Kr](1+br))$. $\zeta = 0.179$ is related to the ratio between the two isotopes of Rb. $[Kr]$ is the number density of Kr. $r = P(N_2)/P(Kr)$ is the ratio of the N_2 pressure $P(N_2)$ to the Kr pressure $P(Kr)$ which is around 15 in our design. The coefficient $b = P_0(N_2)/P_0(Kr) = 1.9$, where $P_0(N_2)$ and $P_0(Kr)$ are the characteristic pressures defined in the reference [33]. We finally could calculate that κ_T is an order smaller than that of the binary spin exchange rate coefficient $\sigma_{se}\bar{v}$. Thus, we could neglect the van der Waals molecule formation-related spin exchange optical pumping.

We know that comagnetometers are also sensitive to rotation velocity. For the ^{83}Kr-based atomic comagnetometer, the nuclear spin is 9/2, and we estimated a very small gyro-magnetic ratio, which means that this comagnetometer is quite sensitive to rotation. When performing movement-related MEG, we can only measure movements without rotation. Further experiments need to be performed to study this topic. The study of ^{83}Kr-based atomic comagnetometers also contributes to the study of gyroscopes.

Relaxation rate, polarization and fundamental sensitivity are important parameters for atomic comagnetometers. Note that ^{83}Kr nuclear spin relaxation is quite complicated, since quadrupolar relaxation at the wall as well as the nuclear spin itself should be considered. Especially, collisions with the wall, desorption energy and collision-related average angle should be studied. The quadrupolar interaction process makes simulations more complicated. Our results give very concrete evidence for future comagnetometer design.

The sensitivity of the comagnetometer is fundamentally determined by Equation (9). It is mainly determined by the total relaxation of Rb R_{tot} and the number of the Rb $n_{Rb}V$ which are used for the magnetic field detection. As we know, we want $n_{Rb}V$ to be as small as possible and thus we can fabricate very small alkali vapor cells. The size of the sensor could be smaller and the spatial resolution of magnetic field measurement could be improved. The total relaxation of Rb R_{tot} is determined by several collision relaxation processes. The merit of using ^{83}Kr is that the spin collision relaxation with ^{83}Kr is quite small compared with that of the Rb-^{129}Xe collision. To further show this clearly, we have conducted a simulation shown in Figure 6. The results show that even with very small volume (0.002 cm^3), the estimated sensitivity of the comagnetometer could reach higher than 10 fT/Hz$^{1/2}$.

In the last part of the paper, we show the alkali fabrication process. It is especially noteworthy that hybrid optical pumping is needed. We give details on how to make a hybrid-pumping vapor cell. We also present the alkali fabrication process with the glass-blowing technique.

With the development of the micro-electro mechanical system (MEMS), the key elements of the comagnetometer such as alkali vapor cell, micro-electrical non-magnetic field heating chips, optical lens and the semiconductor laser diodes can be fabricated into chip-scale. Therefore, the size of the future comagnetometer will be smaller and smaller. As we have calculated, we only need 0.002 cm^3 effective alkali vapor cell space for the comagnetometer. Other elements of the comagnetometer which need to be designed around the vapor cell could also be fabricated to be very small. The comagnetoemter could finally be fabricated in chip-scale.

In comparison, the ^{83}Kr-based atomic comagnetometer could work under 420 K and the compensation bandwidth is around 1 Hz which is similar to that of the ^{21}Ne comagnetometer [47]. However, the ^{21}Ne-based atomic comagnetometer needs to work under around 473 K. Special aluminosilicate glass is required for the cell fabrication. This makes the experiment complicated since it is not easy to obtain the aluminosilicate glass. There are around 3 Amagats ^{21}Ne filled in the vapor cell, while there is only 50 Torr ^{83}Kr filled in the vapor cell. Again, it is easier to fabricate a vapor cell based on ^{83}Kr. The total relaxation of Rb which could determine the sensitivity of the comagnetometer is around 300 s^{-1} in the ^{83}Kr atomic comagnetometer which is only around 1/6 of that of the ^{21}Ne comagnetometer [47]. This would make sure that the sensitivity of the ^{83}Kr-based atomic comagnetometer will be higher.

5. Conclusions

In conclusion, we have theoretically designed a new kind of atomic comagnetometer that is based on ^{83}Kr nuclear spins. Compared with ^{21}Ne-, ^{3}He- and ^{129}Xe-based atomic comagnetometers, the new comagnetometer has the merits of high sensitivity, relatively low temperature and a large compensation field range. We conclude that the ^{83}Kr-based comagnetometer is quite promising for movement-related MEG recording. We give the key parameters for the design of the comagnetometer, such as spin relaxation and polarization. These parameters are highly important for future experiments. Considering that the alkali vapor cell is important for the comagnetometer, we show a method for developing hybrid optical-pumping vapor cells. The study presented in this paper is highly valuable for a future experimental setup.

Author Contributions: Conceptualization, Y.C.; methodology, R.G. and J.W.; investigation, M.Y. and M.Z.; writing—original draft preparation, Y.C.; supervision, L.Z. All authors have read and agreed to the published version of the manuscript.

Funding: This work was supported by National Natural Science Foundation of China under grant number 62103324, China Postdoctoral Science Foundation under grant number 2020M683462, National key research and development program under grant number 2022YFB3203400.

Institutional Review Board Statement: Not applicable.

Informed Consent Statement: Not applicable.

Data Availability Statement: The data underlying the results presented in this paper are not publicly available at this time but may be obtained from the authors upon reasonable request.

Acknowledgments: We want to thank all of these agencies mentioned above for their support.

Conflicts of Interest: The authors declare no conflict of interest.

References

1. Knowlton, R.C.; Shih, J. Magnetoencephalography in Epilepsy. *Epilepsia* **2004**, *45*, 61–71. [CrossRef] [PubMed]
2. Englot, D.J.; Nagarajan, S.S.; Imber, B.S.; Raygor, K.P.; Honma, S.M.; Mizuiri, D.; Mantle, M.; Knowlton, R.C.; Kirsch, H.E.; Chang, E.F. Epileptogenic zone localization using magnetoencephalography predicts seizure freedom in epilepsy surgery. *Epilepsia* **2015**, *56*, 949–958. [CrossRef] [PubMed]
3. Boon, L.I.; Geraedts, V.J.; Hillebrand, A.; Tannemaat, M.R.; Contarino, M.F.; Stam, C.J.; Berendse, H.W. A systematic review of MEG-based studies in Parkinson's disease: The motor system and beyond. *Hum. Brain Mapp.* **2019**, *40*, 2827–2848. [CrossRef]
4. Lin, C.H.; Tierney, T.M.; Holmes, N.; Boto, E.; Leggett, J.; Bestmann, S.; Bowtell, R.; Brookes, M.J.; Barnes, G.R.; Miall, R.C. Using optically pumped magnetometers to measure magnetoencephalographic signals in the human cerebellum. *J. Physiol.* **2019**, *597*, 4309–4324. [CrossRef] [PubMed]
5. López-Sanz, D.; Serrano, N.; Maestú, F. The Role of Magnetoencephalography in the Early Stages of Alzheimer's Disease. *Front. Neurosci.* **2018**, *12*, 572. [CrossRef]
6. Timmermann, L.; Gross, J.; Butz, M.; Kircheis, G.; Haussinger, D.; Schnitzler, A. Pathological oscillatory coupling within the human motor system in different tremor syndromes as revealed by magnetoencephalography. *Neurol. Clin. Neurophysiol. NCN* **2004**, *2004*, 26.
7. Chen, Y.; Zhao, L.; Ma, Y.; Yu, M.; Wang, Y.; Zhang, N.; Wei, K.; Jiang, Z. Spin exchange optically pumped nuclear spin self compensation system for moving magnetoencephalography measurement. *Biomed. Opt. Express* **2022**, *13*, 5937–5951. [CrossRef]
8. Boto, E.; Holmes, N.; Leggett, J.; Roberts, G.; Shah, V.; Meyer, S.S.; Muñoz, L.D.; Mullinger, K.J.; Tierney, T.M.; Bestmann, S. Moving magnetoencephalography towards real-world applications with a wearable system. *Nature* **2018**, *555*, 657. [CrossRef]
9. Kominis, I.; Kornack, T.; Allred, J.; Romalis, M.V. A subfemtotesla multichannel atomic magnetometer. *Nature* **2003**, *422*, 596–599. [CrossRef]
10. Yan, Y.; Lu, J.; Zhang, S.; Lu, F.; Yin, K.; Wang, K.; Zhou, B.; Liu, G. Three-axis closed-loop optically pumped magnetometer operated in the serf regime. *Opt. Express* **2022**, *30*, 18300–18309. Available: https://opg.optica.org/oe/abstract.cfm?URI=oe-30-11-18300 (accessed on 8 September 2023).
11. Xia, H.; Ben-Amar Baranga, A.; Hoffman, D.; Romalis, M. Magnetoencephalography with an atomic magnetometer. *Appl. Phys. Lett.* **2006**, *89*, 211104. [CrossRef]
12. Shah, V.K.; Wakai, R.T. A compact, high performance atomic magnetometer for biomedical applications. *Phys. Med. Biol.* **2013**, *58*, 8153. [CrossRef] [PubMed]
13. Borna, A.; Carter, T.R.; Colombo, A.P.; Jau, Y.Y.; McKay, J.; Weisend, M.; Taulu, S.; Stephen, J.M.; Schwindt, P.D. Non-invasive functional-brain-imaging with an OPM-based magnetoencephalography system. *PLoS ONE* **2020**, *15*, e0227684. [CrossRef]
14. Boto, E.; Seedat, Z.A.; Holmes, N.; Leggett, J.; Hill, R.M.; Roberts, G.; Shah, V.; Fromhold, T.M.; Mullinger, K.J.; Tierney, T.M.; et al. Wearable neuroimaging: Combining and contrasting magnetoencephalography and electroencephalography. *NeuroImage* **2019**, *201*, 116099. [CrossRef]
15. Holmes, N.; Leggett, J.; Boto, E.; Roberts, G.; Hill, R.M.; Tierney, T.M.; Shah, V.; Barnes, G.R.; Brookes, M.J.; Bowtell, R. A bi-planar coil system for nulling background magnetic fields in scalp mounted magnetoencephalography. *Neuroimage* **2018**, *181*, 760–774. [CrossRef]

16. Holmes, N.; Rea, M.; Hill, R.M.; Leggett, J.; Edwards, L.J.; Hobson, P.J.; Boto, E.; Tierney, T.M.; Rier, L.; Rivero, G.R.; et al. Enabling ambulatory movement in wearable magnetoencephalography with matrix coil active magnetic shielding. *NeuroImage* **2023**, *274*, 120157. [CrossRef]
17. Rea, M.; Holmes, N.; Hill, R.M.; Boto, E.; Leggett, J.; Edwards, L.J.; Woolger, D.; Dawson, E.; Shah, V.; Osborne, J.; et al. Precision magnetic field modelling and control for wearable magnetoencephalography. *NeuroImage* **2021**, *241*, 118401. [CrossRef] [PubMed]
18. Mellor, S.; Tierney, T.M.; O'Neill, G.C.; Alexander, N.; Seymour, R.A.; Holmes, N.; López, J.D.; Hill, R.M.; Boto, E.; Rea, M.; et al. Magnetic field mapping and correction for moving OP-MEG. *IEEE Trans. Biomed. Eng.* **2021**, *69*, 528–536. [CrossRef] [PubMed]
19. Tierney, T.M.; Alexander, N.; Mellor, S.; Holmes, N.; Seymour, R.; O'Neill, G.C.; Maguire, E.A.; Barnes, G.R. Modelling optically pumped magnetometer interference in MEG as a spatially homogeneous magnetic field. *NeuroImage* **2021**, *244*, 118484. [CrossRef]
20. Pratt, E.J.; Ledbetter, M.; Jiménez-Martínez, R.; Shapiro, B.; Solon, A.; Iwata, G.I.; Garber, S.; Gormley, J.; Decker, D.; Deladillo, D.; et al. Kernel Flux: A whole-head 432-magnetometer optically-pumped magnetoencephalography (OP-MEG) system for brain activity imaging during natural human experiences. In Proceedings of the SPIE 11700, Optical and Quantum Sensing and Precision Metrology, Online, 6–12 March 2021; Volume 11700, p. 1170032.
21. Ghosh, R.K.; Romalis, M.V. Measurement of spin-exchange and relaxation parameters for polarizing ^{21}Ne with K and Rb. *Phys. Rev. A* **2010**, *81*, 043415. [CrossRef]
22. Romalis, M. Hybrid optical pumping of optically dense alkali-metal vapor without quenching gas. *Phys. Rev. Lett.* **2010**, *105*, 243001. [CrossRef]
23. Brown, J.M.; Smullin, S.J.; Kornack, T.W.; Romalis, M.V. New Limit on Lorentz- and CPT-Violating Neutron Spin Interactions. *Phys. Rev. Lett.* **2010**, *105*, 151604. [CrossRef]
24. Li, R.; Fan, W.; Jiang, L.; Duan, L.; Quan, W.; Fang, J. Rotation sensing using a K-Rb-^{21}Ne comagnetometer. *Phys. Rev. A* **2016**, *94*, 032109. [CrossRef]
25. Fang, J.; Chen, Y.; Zou, S.; Liu, X.; Hu, Z.; Quan, W.; Yuan, H.; Ding, M.; Physics, O. Low frequency magnetic field suppression in an atomic spin co-magnetometer with a large electron magnetic field. *J. Phys. B At. Mol.* **2016**, *49*, 065006. [CrossRef]
26. Zhang, S.B.; Ba, Z.L.; Ning, D.H.; Zhai, N.F.; Lu, Z.T.; Sheng, D. Search for Spin-Dependent Gravitational Interactions at Earth Range. *Phys. Rev. Lett.* **2023**, *130*, 201401. [CrossRef]
27. Ji, W.; Chen, Y.; Fu, C.; Ding, M.; Fang, J.; Xiao, Z.; Wei, K.; Yan, H. New Experimental Limits on Exotic Spin-Spin-Velocity-Dependent Interactions by Using $SmCo_5$ Spin Sources. *Phys. Rev. Lett.* **2018**, *121*, 261803. [CrossRef]
28. Kornack, T.; Ghosh, R.; Romalis, M.V. Nuclear spin gyroscope based on an atomic comagnetometer. *Phys. Rev. Lett.* **2005**, *95*, 230801. [CrossRef]
29. Vasilakis, G.; Brown, J.; Kornack, T.; Romalis, M. Limits on New Long Range Nuclear Spin-Dependent Forces Set with a K-^{3}He Comagnetometer. *Phys. Rev. Lett.* **2009**, *103*, 261801. [CrossRef]
30. Kornack, T.; Romalis, M. Dynamics of two overlapping spin ensembles interacting by spin exchange. *Phys. Rev. Lett.* **2002**, *89*, 253002. [CrossRef]
31. Wu, Z.; Happer, W.; Kitano, M.; Daniels, J. Experimental studies of wall interactions of adsorbed spin-polarized 131xe nuclei. *Phys. Rev. A* **1990**, *42*, 2774. [CrossRef]
32. Butscher, R.; Wäckerle, G.; Mehring, M. Nuclear quadrupole surface interaction of gas phase 83Kr: Comparison with 131Xe. *Chem. Phys. Lett.* **1996**, *249*, 444–450. [CrossRef]
33. Schaefer, S.R.; Cates, G.D.; Happer, W. Determination of spin-exchange parameters between optically pumped rubidium and ^{83}Kr. *Phys. Rev. A* **1990**, *41*, 6063–6070. [CrossRef]
34. Brinkmann, D.; Kuhn, D. Nuclear magnetic relaxation of ^{83}Kr in krypton gas. *Phys. Rev. A* **1980**, *21*, 163–167. [CrossRef]
35. Wu, Z.; Schaefer, S.; Cates, G.D.; Happer, W. Coherent interactions of the polarized nuclear spins of gaseous atoms with the container walls. *Phys. Rev. A* **1988**, *37*, 1161–1175. [CrossRef]
36. Butscher, R.; Wäckerle, G.; Mehring, M. Nuclear quadrupole interaction of highly polarized gas phase 131Xe with a glass surface. *J. Chem. Phys.* **1994**, *100*, 6923–6933. [CrossRef]
37. Smiciklas, M.; Brown, J.M.; Cheuk, L.W.; Smullin, S.J.; Romalis, M.V. New Test of Local Lorentz Invariance Using a ^{21}Ne-Rb-K Comagnetometer. *Phys. Rev. Lett.* **2011**, *107*, 171604. [CrossRef]
38. Walker, T.G. Estimates of spin-exchange parameters for alkali-metal—Noble-Gas Pairs. *Phys. Rev. A* **1989**, *40*, 4959–4964. [CrossRef]
39. Walker, T.G.; Happer, W. Spin-exchange optical pumping of noble-gas nuclei. *Rev. Mod. Phys.* **1997**, *69*, 629. [CrossRef]
40. Fang, J.; Chen, Y.; Lu, Y.; Quan, W.; Zou, S.; Physics, O. Dynamics of Rb and 21Ne spin ensembles interacting by spin exchange with a high Rb magnetic field. *J. Phys. B At. Mol.* **2016**, *49*, 135002. [CrossRef]
41. Allred, J.C.; Lyman, R.N.; Kornack, T.W.; Romalis, M.V. High-Sensitivity Atomic Magnetometer Unaffected by Spin-Exchange Relaxation. *Phys. Rev. Lett.* **2002**, *89*, 130801. [CrossRef]
42. Chen, Y.; Quan, W.; Duan, L.; Lu, Y.; Jiang, L.; Fang, J. Spin-exchange collision mixing of the K and Rb ac Stark shifts. *Phys. Rev. A* **2016**, *94*, 052705. [CrossRef]
43. Chen, W.C.; Gentile, T.R.; Walker, T.G.; Babcock, E. Spin-exchange optical pumping of ^{3}He with Rb-K mixtures and pure K. *Phys. Rev. A* **2007**, *75*, 013416. [CrossRef]
44. Chen, W.C.; Gentile, T.R.; Fu, C.B.; Watson, S.; Jones, G.L.; McIver, J.W.; Rich, D.R. Polarized 3He cell development and application at NIST. *J. Phys. Conf. Ser.* **2011**, *294*, 012003. [CrossRef]

45. Schaefer, S.R.; Cates, G.D.; Chien, T.R.; Gonatas, D.; Happer, W.; Walker, T.G. Frequency shifts of the magnetic-resonance spectrum of mixtures of nuclear spin-polarized noble gases and vapors of spin-polarized alkali-metal atoms. *Phys. Rev. A* **1989**, *39*, 5613–5623. [CrossRef]
46. Cates, G.; Fitzgerald, R.; Barton, A.; Bogorad, P.; Gatzke, M.; Newbury, N.; Saam, B.J.P.R.A. Rb− 129Xe spin-exchange rates due to binary and three-body collisions at high Xe pressures. *Phys. Rev. A* **1992**. *45*, 4631. [CrossRef]
47. Chen, Y.; Quan, W.; Zou, S.; Lu, Y.; Duan, L.; Li, Y.; Zhang, H.; Ding, M.; Fang, J. Spin exchange broadening of magnetic resonance lines in a high-sensitivity rotating K-Rb-^{21}Ne co-magnetometer. *Sci. Rep.* **2016**, *6*, 36547. [CrossRef]

photonics

MDPI

Communication

Nonzero-Order Resonances in Single-Beam Spin-Exchange Relaxation-Free Magnetometers

Kun Wang [1,2], Kaixuan Zhang [2,*], Nuozhou Xu [1,2], Yifan Yan [1,2], Xiaoyu Li [1,2] and Binquan Zhou [1,2]

1 Key Laboratory of Ultra-Weak Magnetic Field Measurement Technology, Ministry of Education, School of Instrumentation and Optoelectronic Engineering, Beihang University, Beijing 100191, China
2 Zhejiang Provincial Key Laboratory of Ultra-Weak Magnetic-Field Space and Applied Technology, Hangzhou Innovation Institute, Beihang University, Hangzhou 310051, China
* Correspondence: zhangkx@shanghaitech.edu.cn

Abstract: Zero-field optically pumped magnetometers operating in the spin-exchange relaxation-free (SERF) regime have been extensively studied, and usually depend on zeroth-order parametric resonance to measure the magnetic field. However, the studies conducted on this topic lack thorough analyses and in-depth discussion of nonzero-order magnetic resonances in single-beam SERF magnetometers. In this paper, we analyzed the nonzero-order resonance, especially the first-order resonance, based on a single-beam SERF magnetometer, and discussed its various applications. A comprehensive theoretical analysis and experiments were conducted with respect to multiple functions, including nonzero finite magnetic field measurements, spin polarization measurement, and in situ coil constant calibration. The results showed that first-order resonance can be utilized for nonzerofinite magnetic field measurements, and the spin polarization of alkali-metal atoms can be determined by measuring the slowing-down factor using the resonance condition. Furthermore, acquiring the first-order resonance point at an equivalent zero pump light power through fitting offers an approach for quick and precise in situ coil constant calibration. This study contributes to the applications of SERF magnetometers in nonzero finite magnetic fields.

Keywords: SERF magnetometers; parametric resonances; slowing-down factor; spin polarization; coil constant calibration

1. Introduction

Optically pumped magnetometers (OPMs) have attracted considerable attention owing to their ultra-high sensitivity, flexible positioning, and cryogenic-free working conditions, and they provide superior advantages over traditional superconducting quantum interference devices [1–3]. They are increasingly used in fundamental physics [4,5], geophysical measurements [6,7], and magnetic imaging of the human body [8,9]. In particular, OPMs operating in the spin-exchange relaxation-free (SERF) regime have undergone rapid development [10,11]. Thanks to their convenient miniaturization, SERF magnetometers with single-beam configuration, employing power detection of transmitted circularly polarized pump light to measure the magnetic field, are the most commonly applied scheme [12–15], and their operation relies on transverse magnetic field modulation and zeroth-order parametric resonance.

The parametric resonance response theory of light-pumped atoms in a modulated magnetic field was dissertated by Cohen-Tannoudji et al. in 1970 [16] and was demonstrated by Slocum et al. in 1973 using a ^{4}He magnetometer [17]; however, they mainly focused on the zeroth-order resonance. Nevertheless, there are still only a few studies on the first-order or higher-order resonances. For instance, Xiao et al. proposed that the first-order resonance of a SERF magnetometer could be employed to calibrate coil constants and simultaneously showed the possibility of measuring a magnetic field under a large field background [18]. For other types of magnetometers, Eklund [19] and Chen et al. [20] employed the resonance order of rubidium magnetization and discussed its application in nuclear magnetic resonance. Jiang et al. developed a rubidium atomic magnetometer that satisfies the first-order resonance to study the heater-induced longitudinal magnetic field [21]. Yang et al. proposed a novel plan for magnetic sensing based on multi-order resonance utilizing an M_x magnetometer, and the measurement sensitivity was below 3 pT/Hz$^{1/2}$ [22]. However, these studies lack comprehensive analyses and detailed discussion of nonzero-order magnetic resonances in single-beam SERF magnetometers; therefore, a comprehensive study on this topic is still desirable.

In this study, we analyzed several phenomena with respect to the nonzero-order resonances of single-beam SERF magnetometers and discussed their applications, including spin polarization measurement and coil constant calibration. We conducted a theoretical analysis, numerical simulation, and experiments to study the parametric resonance phenomena and their applications. First, a nonzero finite magnetic field was measured. A finite magnetic field was acquired by detecting the resonant frequency in combination with the resonance condition. Second, the spin polarization of alkali-metal atoms was measured by measuring the slowing-down factor $q(P)$ based on the resonance condition. Finally, the first-order resonance point at the equivalent zero light power with a definitive known $q(P)$ was determined by fitting; this offered a method to calibrate the coil constants rapidly and precisely. Moreover, the sensitivity of the magnetic field measurement reached 54 fT/Hz$^{1/2}$.

2. Principles

OPMs are employed to measure the magnetic field by detecting the time evolution of atomic spin polarization created by optical pumping. When the spin-exchange rate is significantly higher than the Larmor precession frequency, the Bloch equation derived from the density matrix equations can be adopted to describe the behavior of the atomic spin polarization vector **P** [23] as follows:

$$\frac{d\mathbf{P}}{dt} = \frac{1}{q(P)}\left[\gamma^e \mathbf{B} \times \mathbf{P} + R_{\text{op}}(s\hat{z} - \mathbf{P}) - R_{\text{rel}}\mathbf{P}\right] \tag{1}$$

where P is the magnitude of the atomic spin polarization vector, $\mathbf{P} = \left[P_x, P_y, P_z\right]^T$; and $q(P)$ is the nuclear slowing-down factor as a function of P [24]. In particular, for ^{87}Rb atoms ($I = 3/2$) the expression of $q(P)$ is given as

$$q(P) = \frac{6 + 2P^2}{1 + P^2} \tag{2}$$

which indicates that the amplitude of spin polarization can be obtained by measuring $q(P)$. In addition, $\gamma^e \approx 2\pi \times 28$ Hz/nT is the gyromagnetic ratio of the electron, **B** is the magnetic field vector expressed as $\mathbf{B} = \left[B_x, B_y, B_z\right]^T$, R_{op} is the optical pumping rate, R_{rel} is the spin-relaxation rate, and s is the degree of circular polarization of the pump light whose direction propagates along the z-axis; for circularly polarized light, this is s = ±1.

Specifically, for single-beam OPMs, the transmitted light intensity of a circularly polarized laser beam which only reflects the longitudinal polarization P_z can act as a magnetic field information medium [25–27]. The relationship between the output of the magnetometer R_{out} and P_z is given as [25]

$$R_{out} = R_0 \cdot \exp[-\text{OD}(\nu)(1 - P_z)] \tag{3}$$

where R_0 is the original signal before the vapor cell; $\text{OD}(\nu) = n_\nu L\sigma(\nu)$ is the optical depth of the vapor cell; n_ν is the number density of the alkali-metal vapor; $\sigma(\nu)$ is the absorption cross-section as a function of light frequency ν; and L is the length of the vapor cell. Thus, the response of P_z to a magnetic field is the subject of concern. In single-beam SERF magnetometers, a bias magnetic field B_{x0} and a modulation field $B_m\cos(\omega t)$ were applied along the x-axis, as shown in Figure 1a. After magnetic compensation, the magnetic field along the three axes was described as $\mathbf{B} = [B_{x0} + B_m\cos(\omega t), 0, 0]^T$. From Equation (1), we know that $P_x(t)$ approaches zero, and $\mathbf{P}$ oscillates and evolves only in the y-z plane. Therefore, we define polarization as $P_+ = P_z + iP_y$, which follows the evolution

$$\frac{dP_+}{dt} = \frac{1}{q(P)}\{-i\gamma^e[B_{x0} + B_m\cos(\omega t)]P_+ - (R_{\mathrm{op}} + R_{\mathrm{rel}})P_+ + R_{\mathrm{op}}\} \tag{4}$$

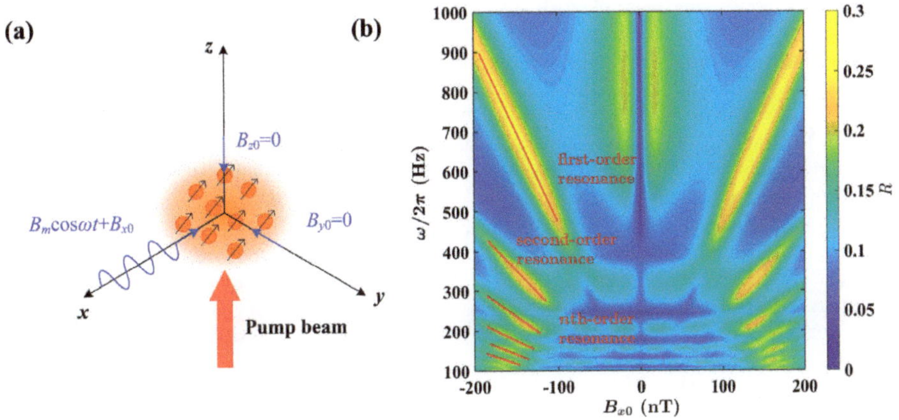

Figure 1. (**a**) Schematic of our single-beam SERF magnetometer with the modulated magnetic field along the x-axis. (**b**) Simulation results of the magnetometer response R under different $\omega/2\pi$ and B_{x0}, with $B_m = 200$ nT.

After employing the Jacobi-Anger expansion and mathematical derivation, the analytical solution to P_z can be obtained as [16,17,19]

$$\begin{aligned} P_z(t) &= \sum_{n=-\infty}^{+\infty}\sum_{k=-\infty}^{+\infty}\frac{R_{\mathrm{op}}J_n(u)J_{n+k}(u)}{(R_{\mathrm{op}}+R_{\mathrm{rel}})^2+(\gamma^e B_{x0}+nq\omega)^2}[(R_{\mathrm{op}} + R_{\mathrm{rel}})\cos(k\omega t) \\ &+ (\gamma^e B_{x0} + nq\omega)\sin(k\omega t)] \end{aligned} \tag{5}$$

where $J_n(u)$ is the nth-order Bessel function of the first kind; and $u = \gamma^e B_m/(q\omega)$ is the modulation index. Here, n denotes the resonance order, with each order appearing at the magnetic field offset satisfying $\gamma^e B_{x0} + nq\omega = 0$.

To eliminate low-frequency 1/f noise, the lock-in detection system for the first harmonic ω, the dominant term of the spectrum, was employed. The longitudinal polarization embodied in the demodulated signal is $P_{z-de}(t)$, which is expressed as follows:

$$\begin{aligned} P_{z-de}(t) = &\sum_{n=-\infty}^{+\infty}\frac{R_{\mathrm{op}}J_n(u)}{(R_{\mathrm{op}}+R_{\mathrm{rel}})^2+(\gamma^e B_{x0}+nq\omega)^2}\{(R_{\mathrm{op}} + R_{\mathrm{rel}})[J_{n+1}(u) + J_{n-1}(u)]\cos(\omega t) \\ &+ (\gamma^e B_{x0} + nq\omega)[J_{n+1}(u) - J_{n-1}(u)]\sin(\omega t)\} \end{aligned} \tag{6}$$

The demodulated in-phase component X_n and out-of-phase component Y_n for the nth-order resonance are

$$X_n = J_n(u)[J_{n+1}(u) + J_{n-1}(u)]\frac{R_{\mathrm{op}}(R_{\mathrm{op}} + R_{\mathrm{rel}})}{(R_{\mathrm{op}} + R_{\mathrm{rel}})^2 + (\gamma^e B_{x0} + nq\omega)^2} \tag{7}$$

$$Y_n = J_n(u)[J_{n+1}(u) - J_{n-1}(u)]\frac{R_{op}(\gamma^e B_{x0} + nq\omega)}{(R_{op} + R_{rel})^2 + (\gamma^e B_{x0} + nq\omega)^2} \tag{8}$$

The amplitude component R_n depends both on X_n and Y_n:

$$R_n = \sqrt{X_n^2 + Y_n^2} \tag{9}$$

Here, X_n and Y_n are closely related to the phase of the demodulation process, and the phase varies when the sweeping magnetic field is offset. Consequently, R_n was selected as observable in our experiment. The Y_n component presented a dispersion relation with B_{x0} near the nth resonance point $B_{xn} = -nq\omega/\gamma^e$, which can be utilized for magnetic field measurement. The total in-phase component X, total out-of-phase component Y, and total amplitude component R are expressed as $X = \sum_{n=-\infty}^{n=+\infty} X_n$, $Y = \sum_{n=-\infty}^{n=+\infty} Y_n$, and $R = \sum_{n=-\infty}^{n=+\infty} R_n$, respectively.

The analysis of each order of resonance was based on the response amplitude R of the magnetometer, owing to its insensitivity to the demodulation phase. We performed a simulation to calculate the magnetometer response R under different $\omega/2\pi$ and B_{x0}, with typical experimental parameters of $R_{op} = R_{rel} = 250\ \text{s}^{-1}$ and $B_m = 200$ nT. The simulation covered a span range of $\omega/2\pi$ from 100 Hz to 1000 Hz and a span range of B_{x0} from -200 nT to 200 nT, as depicted in Figure 1b. From the simulation, we found that the resonance point B_{x0} of each order was linear to $\omega/2\pi$ under resonance conditions, whereas the interval between each order increased when $\omega/2\pi$ grew. Furthermore, the attenuation of the modulation index u, corresponding to the growth of $\frac{\omega}{2\pi}$, results in the gradual dominance of the lower-order resonance, indicating that u must be controlled in the appropriate range to maintain the predominance of the first-order resonance. In other words, the higher-order resonance response of the magnetometer ($n > 1$) only emerges markedly with lower $\omega/2\pi$ (<350 Hz, shown in Figure 1b), which restricts the performance of the lock-in system and further limits the sensitivity of measurement. Moreover, Figure 1b shows that the higher-order resonance response ($n > 1$) appears at an absolute value of magnetic offset larger than 100 nT, which can induce strong relaxation of alkali atoms. Hence, in this study, our attention was mostly paid to the first-order resonance response.

In addition, by focusing on the amplitude of the demodulated signal at each resonance order, we found that if $n \neq 0$ then $Y_n = 0$. By contrast, when $n = 0$, $X_n = 0$. Consequently, the demodulated signal at the nth-order resonance is given as

$$R_n^{res} = \begin{cases} \frac{2J_0(u)J_1(u)R_{op}\gamma^e B_{x0}}{(R_{op}+R_{rel})^2+(\gamma^e B_{x0})^2} = 0 & n = 0 \\ \frac{2nJ_n^2(u)}{u} \cdot \frac{R_{op}}{R_{op}+R_{rel}} & n \neq 0 \end{cases} \tag{10}$$

and the linear region of R_n^{res} could be utilized to measure the low-frequency alternating magnetic field [22].

We revealed that the magnitude of spin polarization at the first-order resonance point is constant and time-independent. At the first-order resonance point, by substituting $n = 1$ ($n = -1$ is similar) and $\gamma^e B_{x0} + q\omega = 0$ conditions for the corresponding expressions in Equation (5), we obtain

$$\begin{aligned} P_z(t) &= \frac{R_{op}J_1(u)}{R_{op}+R_{rel}} \sum_{k=-\infty}^{+\infty} J_{k+1}(u)\cos(k\omega t) \\ &= \frac{R_{op}J_1(u)}{R_{op}+R_{rel}} \{\sin(\omega t)\sin[u\sin(\omega t)] + \cos(\omega t)\cos[u\sin(\omega t)]\} \end{aligned} \tag{11}$$

Similar equations can be derived for $P_y(t)$ utilizing the same method as

$$\begin{aligned} P_y(t) &= \frac{R_{op}J_1(u)}{R_{op}+R_{rel}} \sum_{k=-\infty}^{+\infty} J_{k+1}(u)\sin(k\omega t) \\ &= \frac{R_{op}J_1(u)}{R_{op}+R_{rel}} \{\cos(\omega t)\sin[u\sin(\omega t)] - \sin(\omega t)\cos[u\sin(\omega t)]\} \end{aligned} \tag{12}$$

As $P_x(t)$ approaches zero, the amplitude of total spin polarization is given as

$$P = \sqrt{[P_z(t)]^2 + [P_y(t)]^2} = \frac{R_{op}J_1(u)}{R_{op} + R_{rel}} \tag{13}$$

Here, we assumed that R_{op}, R_{rel}, and $J_1(u)$ are invariant during the experiment. The resulting stable magnitude of P can be applied to derive the fit function in the following Sections 4.2 and 4.3.

3. Experimental Setup and Procedure

The experimental setup for measuring the magnetic field, based on the first-order parametric resonances of our single-beam SERF magnetometer, is illustrated in Figure 2. A cubic vapor cell made of borosilicate glass, with an internal size of 8 mm × 8 mm × 8 mm, was filled with a droplet of ^{87}Rb and approximately 600 Torr N_2. The vapor cell was electrically heated to 433 K using an alternating current at 200 kHz.

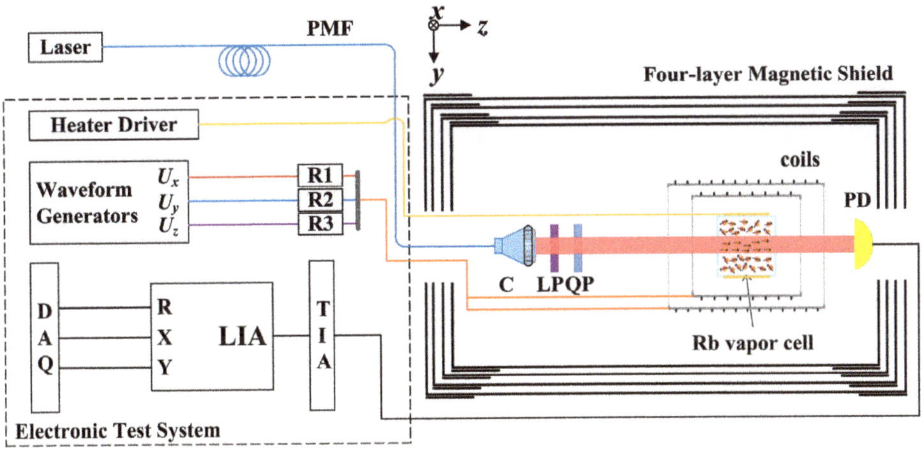

Figure 2. Experimental setup. PMF: polarization maintaining fiber; C: collimating lens; LP: linear polarizer; QP: quarter-wave plate; PD: photodiode; TIA: transimpedance amplifier; LIA: lock-in amplifiers; DAQ: data-acquisition; R1, R2, and R3: resistors.

The laser beam generated by a distributed-feedback laser was first transmitted to the OPM via a polarization-maintaining fiber, and then was transformed into circularly polarized light using a quarter-wave plate, which finally illuminated the vapor cell; this acts as both an optical pumping and probing light. The wavelength of the laser beam was set to approximately 794.98 nm, near the ^{87}Rb D1 line. The light transmitted through the vapor cell was sensed and converted into a current signal by the photodiode, then finally transformed into a voltage signal by the trans-impedance amplifier (PDA200C; Thorlabs, Newton, MA, USA).

The voltage signal was then sent to the electronic test system for signal processing. In the electronic system, a lock-in amplifier (MFLI; Zurich Instruments, Zürich, Switzerland) was used to demodulate the magnetometer signal. Finally, all signals were acquired using the data-acquisition system (PXIe-4464; National Instruments, Austin, TX, USA).

A four-layer μ-metal magnetic shield was utilized to ensure a near-zero magnetic field environment for the OPM. In addition, a group of triaxial coils inside the shield was adopted for further active magnetic field compensation and generation of modulated magnetic fields. The coil group comprised a nested saddle coil [28] for the radial magnetic field and a Lee-Whiting coil [29] for the axial magnetic field; these were both driven by waveform generators (33522B; Keysight, Santa Rosa, United States) via selected resistors.

The experiment primarily comprised three steps. First, we demonstrated the measurement of the non-zero finite magnetic field based on the first-order resonance of the magnetometer. The measurement of the non-zero finite field was realized by detecting the resonance frequency, according to $B_{\text{un}} = nq(P)\omega/\gamma^e$. Second, spin polarization measurement was implemented by measuring the slowing-down factor, and this process was staged with different modulated magnetic field amplitudes. Finally, we performed a precise coil constant calibration by fitting the first-order resonance points under different pump light powers and acquiring the point at equivalent zero pump light powers.

4. Results and Discussions

4.1. Nonzero Finite Magnetic Field Measurement

In this subsection, a method for measuring the non-zero finite magnetic field based on the first-order resonance is proposed and demonstrated. The non-zero finite field was measured by detecting the resonant frequency. Before sweeping the modulated magnetic field frequency, we first swept the magnetic field offset to acquire the slowing-down factor q at each first-order resonance point. Specifically, the magnetic field offset was swept under different modulated magnetic field frequencies $\omega/2\pi$, from 500 Hz to 700 Hz, with a fixed modulated amplitude $B_m = 360$ nT. At each turn the first-order resonance points were recorded, as depicted in Figure 3a.

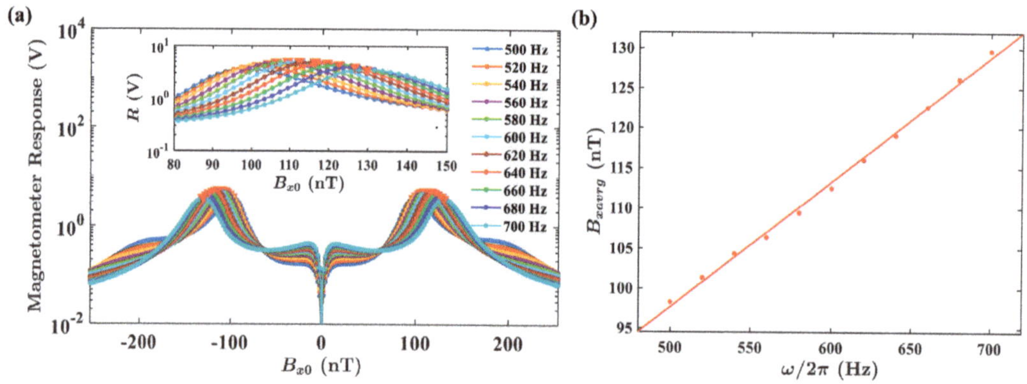

Figure 3. (**a**) Magnetometer responses when sweeping the magnetic field offset B_{x0} under different modulated magnetic field frequencies ranging from 500 Hz to 700 Hz. Each peak of R at the first-order resonance is marked with a red inverted triangle and recorded. (**b**) Each first-order resonance point B_{xavrg} (red solid circle) with different modulated magnetic frequencies $\omega/2\pi$. The fit value is presented as a red solid line.

Two first-order resonance points, corresponding to B_{x1} with $n = 1$ and B_{x-1} with $n = -1$, were extracted, averaged as B_{xavrg}, and plotted with the corresponding $\omega/2\pi$ in Figure 3b, wherein the fit value was obtained based on the proportional relationship between B_{xavrg} and $\omega/2\pi$. In addition, the derived value of $q(P)$, which is called q_{fit}, was further calculated for each fit value of B_{xavrg} (the red solid line in Figure 3b). Then, we actively employed different magnetic field offsets along the same x-axis, each corresponding to B_{xavrg}. Simultaneously, we swept $\omega/2\pi$ from 500 Hz to 700 Hz, with the same fixed modulated amplitude $B_m = 360$ nT as that used when sweeping the magnetic field offset. The magnetometer response at each sweeping turn is recorded in Figure 4a, and its peak resonant frequency $\omega_{peak}/2\pi$ is marked with red inverted triangles.

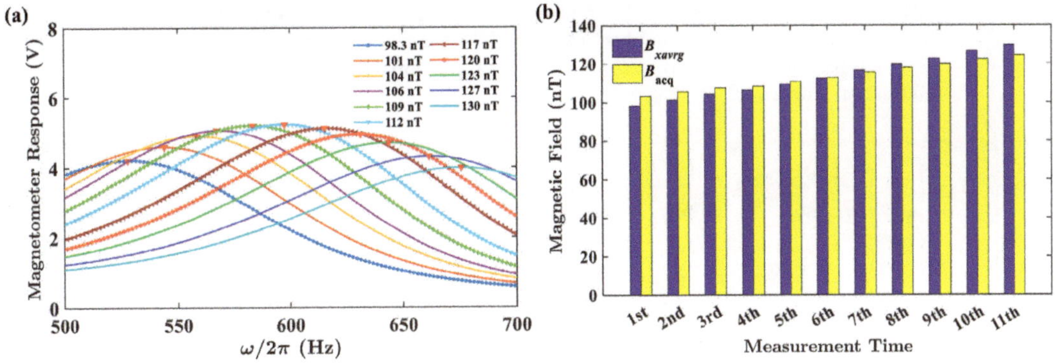

Figure 4. (**a**) Dependence of the magnetometer response on $\omega/2\pi$ under different magnetic field offsets B_{xavrg}. Resonance frequency peaks are marked with red inverted triangles. (**b**) Comparison between the active applied magnetic field B_{xavrg} and B_{acq} acquired from (**a**). The label of horizontal axis "Measurement Time" represents each measurement with different B_{xavrg}.

The measurement results for different finite field offsets are summarized in Figure 4b. The B_{acq} represents the magnetic field acquired by detecting the resonance frequency $\omega_{peak}/2\pi$, i.e., $B_{\text{acq}} = q_{fit}\omega_{peak}/\gamma^e$. The label of horizontal axis "Measurement Time" represents each measurement with different B_{xavrg}, as in Figure 4a, and "1st" and "11th" correspond to "B_{xavrg}= 98.3 nT" and "B_{xavrg}= 130 nT", respectively. The B_{acq} was close to B_{xavrg}, i.e., within a 5% error range, which verified our non-zero finite measurement method based on the first-order resonance. In addition, the difference between B_{acq} and B_{xavrg} may result from the following factors: the residual magnetic field affected the detection of peaks when sweeping B_{x0} and caused the inequality between the actively applied magnetic field B_{act} and B_{xavrg}; the comprehensive effect of magnetic offset and modulated magnetic field induced the variation in the effective gyromagnetic ratio; and the nuclear slowing-down factor [30].

4.2. Spin Polarization Measurement

The spin polarization of ^{87}Rb atoms was obtained by obtaining $q(P)$, which can be measured by determining the first-order resonance point. For instance, we swept the magnetic field offset along the x-axis with different B_m values, ranging from 100 nT peak-peak value to 260 nT peak-peak value, with a $\omega_m/2\pi$ fixed at 600 Hz. The magnetometer response for each sweep is plotted in Figure 5a. The first-order resonance points were extracted and recorded, and each corresponding slowing-down factor (spin polarization) was calculated and plotted in Figure 5b to present the relationship between the slowing-down factor (spin polarization) and B_m. Increasing the modulated magnetic field amplitude B_m leads to a larger relaxation rate R_{rel}.

By utilizing the spin polarization amplitude expression in Equation (13), we determined the relationship between $q(P)$ and B_m. The R_{rel} is composed of $k_{\text{SE}}\gamma_e^2 B_m^2$, related to the modulation amplitude B_m, and R_{rel2}, which is not directly related to B_m. Here, k_{SE} is the proportional factor of R_{rel} related to $\gamma_e^2 B_m^2$. For ^{87}Rb atoms, according to Equation (13), $q(P)$ can be expressed with respect to B_m as follows:

$$q(P) \propto \frac{6\left(R_{\text{op}} + k_{\text{SE}}\gamma_e^2 B_m^2 + R_{\text{rel2}}\right)^2 + 2J_1^2(u)R_{\text{op}}^2}{\left(R_{\text{op}} + k_{\text{SE}}\gamma_e^2 B_m^2 + R_{\text{rel2}}\right)^2 + J_1^2(u)R_{\text{op}}^2} \tag{14}$$

The fit of the experimental data was performed according to Equation (14), and the results indicate that the coefficient of determination r^2 of the fit is 0.993. From Figure 5b, we observed that the larger the modulated magnetic field amplitude, the lower the spin polarization in the stable state. This finding is reasonable and consistent with the theories proposed by Shah et al. [31] and Yan et al. [32], for that the relaxation rate of alkali atoms caused by a modulated magnetic field is proportional to the squared value of the modulation amplitude.

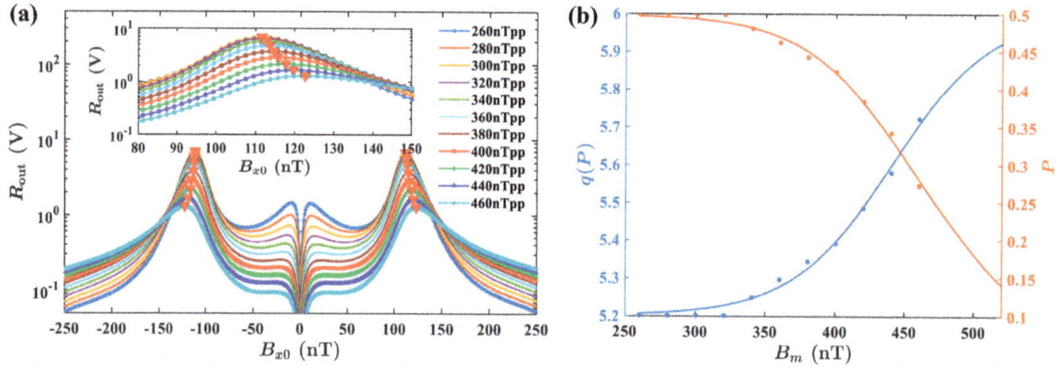

Figure 5. (**a**) Magnetometer response when sweeping the magnetic field offset along the *x*-axis B_{x0} with different B_m values ranging from 260 nT peak-peak value to 460 nT peak-peak value, and the $\omega_m/2\pi$ fixed at 600 Hz. The peak of each first-order resonance is marked with red inverted triangles and recorded. (**b**) The slowing-down factor $q(P)$ and spin polarization P when different B_m values are employed, wherein each record is extracted from the data depicted in subfigure (**a**). The solid circle points represent the experimental data, whereas each solid curve represents the fit curve according to Equation (14).

4.3. In Situ Coil Constant Calibration

In the experiment, the actual value we set when applying a magnetic field was voltage U_{x0}, and the coil constant along the *x*-axis was $k_x = B_{x0}R_x/U_{x0}$. In this equation, R_x is the known resistor value, and the certain value of B_{x0}, i.e., B_{x1}, can be acquired through the resonance condition $\gamma^e B_{x1} + q\omega_{x1} = 0$. The coil constant along the *y*-axis showed a similar relationship. However, simply utilizing the resonance condition to calibrate the coil constant may result in an error caused by the undetermined value of $q(P)$. Our method provides precise calibration based on combining the first-order resonance condition with the definitive value of $q(P)$ obtained by fitting.

Deriving the relationship between ω_{x1} and I_{pump} helps fitting for ω_{x1} when $I_{pump} \to 0$, i.e., $R_{\rm op} \to 0$ causes atoms unpumped with zero P and definitive $q(P) \to 6$ for ^{87}Rb atoms. At the limit of $I_{pump} \to 0$, the magnetic field at the first-order resonance is known to be $B_{x1} = 6\omega_{x1}/\gamma^e$. In this manner, the coil constants can be precisely calibrated as $k_x = B_{x1}R_x/U_{x0}$.

We swept $\omega/2\pi$, from 200 Hz to 1000 Hz, with an invariable B_{x1}. The magnetometer response is illustrated in Figure 6a. At each sweeping turn, the pump light power I_{pump} was set as a different value ranging from 0.40 mW to 2.40 mW, with a step size of 0.20 mW, and the peak of each first-order resonance is marked in Figure 6a and recorded. The other experimental parameters were consistent with simulation parameters.

Subsequently, we listed the first-order resonance points (expressed as the modulation frequency $\omega_{x1}/2\pi$), and plotted Figure 6b to denote the relationship between $\omega_{x1}/2\pi$ and I_{pump}. Determining the resonance in the modulation frequency $\omega_{x1}/2\pi$ is better than reading from the resonance when sweeping the magnetic field offset B_{x0}, due to the error caused by the residual magnetic fields.

Based on Equation (13), the relationship between $\omega_{x1}/2\pi$ and I_{pump} can be derived. $R_{\rm op}$ is approximately proportional to I_{pump}, and $R_{\rm op} \approx k_{\rm op}I_{pump}$. From the first-order resonance condition $\gamma^e B_{x0} + q\omega = 0$, the proportional relationship for $\omega_{x1}/2\pi$ is $\omega_{x1}/2\pi \propto 1/q(P)$. Thereby, for ^{87}Rb atoms the expression can be derived as

$$\omega_{x1}/2\pi \propto \frac{\left(k_{op}I_{pump} + R_{rel}\right)^2 + J_1^2(u)k_{op}^2 I_{pump}^2}{6\left(k_{op}I_{pump} + R_{rel}\right)^2 + 2J_1^2(u)k_{op}^2 I_{pump}^2} \tag{15}$$

Equation (15) is adopted as the fit function between $\omega_{x1}/2\pi$ and I_{pump}.

We employed Equation (15) to fit the data in Figure 6b, and the fit value of ω_{x1} for $I_{pump} \to 0$ with $r^2 = 0.985$ can be obtained as $\omega_{x1\_0} = 517.0$ Hz. Consequently, we showed that $k_x = B_{x1}R_x/U_{x0}$ =14.54 nT/mA, and the coil constant along the same x-axis measured with a flux-gate magnetometer is 14.50 nT/mA, as shown in Figure 7a. This is regarded as the benchmark of the measured result of k_x herein. The coil constants k_x measured through these two methods, with a relative error of 0.29%, are almost similar. In addition, by employing a similar method, the coil constant along the y-axis was acquired as $k_y = B_{y1}R_y/U_{y0}$ =14.53 nT/mA.

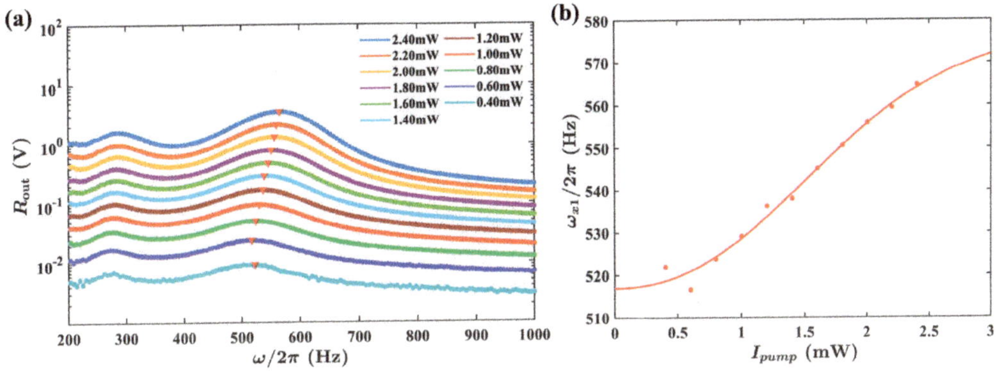

Figure 6. (**a**) Magnetometer response to different frequencies of the modulated magnetic field and with different pump light powers I_{pump} (plotted using different colors). Each peak of magnetometer response of R at the first-order resonance is marked with a red inverted triangle and recorded. (**b**) The plotted $\omega_{x1}/2\pi$ under different I_{pump}, wherein each record is extracted from the data depicted in (**a**). This plot indicates the relationship between $\omega_{x1}/2\pi$ and I_{pump}, and, moreover, the fit value of ω_{x1} for $I_{pump} \to 0$ can be acquired.

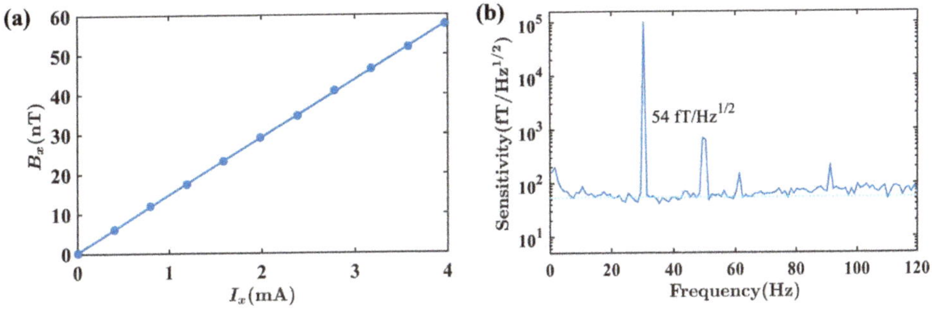

Figure 7. (**a**) The measurement of magnetic field generated by the same coil as that in Section 4.3 with different coil currents, and using a flux-gate magnetometer, for comparison. (**b**) Sensitivity of the magnetic field measurement based on the first-order resonance of single-beam SERF magnetometers. The measurement sensitivity is evaluated by applying the calibration magnetic field signal to the magnetometer along with the modulated magnetic field. The sensitivity of magnetic field measurement is 54 fT/Hz$^{1/2}$.

Furthermore, we evaluated the sensitivity of the magnetic field measurement based on the first-order resonance of a single-beam SERF magnetometer. The measurement sensitivity was analyzed by acquiring the noise spectrum of the output signal of the OPM. For direct measurement, a 100 pT_{rms} magnetic calibration signal at 30.5 Hz and the magnetic field offset corresponding to the first-order resonance pointwere employed along the sensitive axis (x-axis in the experiment). Subsequently, the voltage output signal was acquired and collected for 60 s, and a noise spectral analysis was conducted to determine the sensitivities of the magnetometer. The sensitivity results are given in Figure 7b, where the maximum peaks in each noise spectrum represent the calibration signals. The sensitivity of magnetic field measurement was 54 $fT/Hz^{1/2}$.

5. Conclusions

In this study, we assessed the nonzero-order parametric resonances of alkali-metal atoms in a single-beam SERF magnetometer and primarily focused on the first-order resonance. Based on the first-order resonance, not only can the nonzerofinite magnetic field be measured, but the spin polarization of alkali-metal atoms can also be determined by measuring the slowing-down factor. Moreover, precise calibration of the coil constants can be achieved by acquiring the first-order resonance points under a fitted equivalent zero light power (with a definitive $q(P)$ for certain types of alkali-metal atoms). Our study summarizes parametric resonance, and the proposed method has the potential to function as a component of the systematic analysis of single-beam SERF magnetometers. Future studies may focus on the comprehensive response of SERF magnetometers at different harmonic and resonance orders.

Author Contributions: Conceptualization, K.W. and K.Z.; methodology, K.W.; software, N.X.; validation, K.W., X.L. and K.Z.; formal analysis, K.W.; investigation, Y.Y.; resources, K.Z.; data curation, K.W.; writing—original draft preparation, K.W.; writing—review and editing, K.Z.; visualization, K.W.; supervision, B.Z.; project administration, K.Z.; funding acquisition, B.Z. All authors have read and agreed to the published version of the manuscript.

Funding: This study was supported by the National Natural Science Foundation of China under Grant (No. 61903013).

Institutional Review Board Statement: Not applicable.

Informed Consent Statement: Not applicable.

Data Availability Statement: The data presented in this study are available upon reasonable request from the corresponding author.

Conflicts of Interest: The authors declare no conflict of interest.

References

1. Kominis, I.K.; Kornack, T.W.; Allred, J.C.; Romalis, M.V. A subfemtotesla multichannel atomic magnetometer. *Nature* **2003**, *422*, 596–599. [CrossRef]
2. Wakai, R.T. The atomic magnetometer: A new era in biomagnetism. *AIP Conf. Proc.* **2014**, *1626*, 46–54. [CrossRef]
3. Boto, E.; Holmes, N.; Leggett, J.; Roberts, G.; Shah, V.; Meyer, S.S.; Muñoz, L.D.; Mullinger, K.J.; Tierney, T.M.; Bestmann, S.; et al. Moving magnetoencephalography towards real-world applications with a wearable system. *Nature* **2018**, *555*, 657–661. [CrossRef] [PubMed]
4. Abel, C.; Afach, S.; Ayres, N.J.; Baker, C.A.; Ban, G.; Bison, G.; Bodek, K.; Bondar, V.; Burghoff, M.; Chanel, E.; et al. Measurement of the Permanent Electric Dipole Moment of the Neutron. *Phys. Rev. Lett.* **2020**, *124*, 081803. [CrossRef]
5. Kim, Y.J.; Chu, P.-H.; Savukov, I.; Newman, S. Experimental limit on an exotic parity-odd spin- and velocity-dependent interaction using an optically polarized vapor. *Nat. Commun.* **2019**, *10*, 2245. [CrossRef] [PubMed]
6. Dang, H.B.; Maloof, A.C.; Romalis, M.V. Ultrahigh sensitivity magnetic field and magnetization measurements with an atomic magnetometer. *Appl. Phys. Lett.* **2010**, *97*, 151110. [CrossRef]
7. Higbie, J.M.; Rochester, S.M.; Patton, B.; Holzlöhner, R.; Calia, D.B.; Budker, D. Magnetometry with Mesospheric Sodium. *Proc. Nat. Acad. Sci. USA* **2011**, *108*, 3522–3525. [CrossRef]
8. Iivanainen, J.; Zetter, R.; Parkkonen, L. Potential of on-scalp MEG: Robust detection of human visual gamma-band responses. *Hum. Brain Mapp.* **2020**, *41*, 150–161. [CrossRef] [PubMed]

9. Boto, E.; Shah, V.; Hill, R.M.; Rhodes, N.; Osborne, J.; Doyle, C.; Holmes, N.; Rea, M.; Leggett, J.; Bowtell, R.; et al. Triaxial detection of the neuromagnetic field using optically-pumped magnetometry: Feasibility and application in children. *NeuroImage* **2022**, *252*, 119027. [CrossRef]
10. Li, Z.; Wakai, R.T.; Walker, T.G. Parametric modulation of an atomic magnetometer. *Appl. Phys. Lett.* **2006**, *89*, 134105. [CrossRef]
11. Hu, Y.; Hu, Z.; Liu, X.; Li, Y.; Zhang, J.; Yao, H.; Ding, M. Reduction of far off-resonance laser frequency drifts based on the second harmonic of electro-optic modulator detection in the optically pumped magnetometer. *Appl. Opt.* **2017**, *56*, 5927–5932. [CrossRef]
12. Osborne, J.; Orton, J.; Alem, O.; Shah, V. Fully integrated, standalone zero field optically pumped magnetometer for biomagnetism. In *Proceedings of the Proceedings SPIE 10548, Steep Dispersion Engineering and Opto-Atomic Precision Metrology XI*; Shahriar, S.M., Scheuer, J., Eds.; SPIE: San Francisco, CA, USA, 2018; p. 10548. [CrossRef]
13. Sheng, D.; Perry, A.R.; Krzyzewski, S.P.; Geller, S.; Kitching, J.; Knappe, S. A microfabricated optically-pumped magnetic gradiometer. *Appl. Phys. Lett.* **2017**, *110*, 031106. [CrossRef] [PubMed]
14. Wang, J.; Fan, W.; Yin, K.; Yan, Y.; Zhou, B.; Song, X. Combined effect of pump-light intensity and modulation field on the performance of optically pumped magnetometers under zero-field parametric modulation. *Phys. Rev. A* **2020**, *101*, 053427. [CrossRef]
15. Wang, K.; Zhang, K.; Zhou, B.; Lu, F.; Zhang, S.; Yan, Y.; Wang, W.; Lu, J. Triaxial closed-loop measurement based on a single-beam zero-field optically pumped magnetometer. *Front. Phys.* **2022**, *10*, 1059487. [CrossRef]
16. Cohen-Tannoudji, C.; Dupont-Roc, J.; Haroche, S.; Laloë, F. Diverses résonances de croisement de niveaux sur des atomes pompés optiquement en champ nul. I. Théorie. *Rev. Phys. Appl. (Paris)* **1970**, *5*, 95–101. [CrossRef]
17. Slocum, R.E.; Marton, B.I. Measurement of Weak Magnetic Fields Using Zero-Field Parametric Resonance in Optically Pumped He4. *IEEE Trans. Magn.* **1973**, *9*, 221–226. [CrossRef]
18. Xiao, W.; Wang, H.; Zhang, X.; Wu, Y.; Wu, T.; Chen, J.; Peng, X.; Guo, H. In Situ Calibration of Magnetic Field Coils Using Parametric Resonance in Optically-pumped Magnetometers. In Proceedings of the 2021 Joint Conference of the European Frequency and Time Forum and IEEE International Frequency Control Symposium (EFTF/IFCS), Gainesville, FL, USA, 7–17 July 2021; pp. 1–3. [CrossRef]
19. Eklund, E.J. Microgyroscope Based on Spin-Polarized Nuclei. PhD Thesis, University of California, Irvine, CA, USA, 2008.
20. Chen, C.; Jiang, Q.; Wang, Z.; Zhang, Y.; Luo, H.; Yang, K. A non-interference method for measurement of transverse relaxation time of the alkali metal magnetometer in nuclear magnetic resonance oscillator. *AIP Adv.* **2020**, *10*, 065303. [CrossRef]
21. Jiang, Q.; Luo, H.; Zhan, X.; Zhang, Y.; Yang, K.; Wang, Z. Avoiding the impact of the heater-induced longitudinal field on atomic magnetometers. *J. Appl. Phys.* **2018**, *124*, 244501. [CrossRef]
22. Yang, H.; Wang, Q.; Zhao, B.; Li, L.; Zhai, Y.; Han, B.; Tang, F. Magnetic field sensing based on multi-order resonances of atomic spins. *Opt. Express* **2022**, *30*, 6618. [CrossRef]
23. Ledbetter, M.P.; Savukov, I.M.; Acosta, V.M.; Budker, D.; Romalis, M.V. Spin-exchange-relaxation-free magnetometry with Cs vapor. *Phys. Rev. A* **2008**, *77*, 033408. [CrossRef]
24. Appelt, S.; Baranga, A.B.-A.; Erickson, C.J.; Romalis, M.V.; Young, A.R.; Happer, W. Theory of spin-exchange optical pumping of ^{3}He and ^{129}Xe. *Phys. Rev. A* **1998**, *58*, 1412–1439. [CrossRef]
25. Dong, H.F.; Fang, J.C.; Zhou, B.Q.; Tang, X.B.; Qin, J. Three-dimensional atomic magnetometry. *Eur. Phys. J. Appl. Phys.* **2012**, *57*, 21004. [CrossRef]
26. Seltzer, S.J. Developments in Alkali-Metal Atomic Magnetometry. PhD Thesis, Princeton University, Princeton, NJ, USA, 2008.
27. Chen, Y.; Zhao, L.; Zhang, N.; Yu, M.; Ma, Y.; Han, X.; Zhao, M.; Lin, Q.; Yang, P.; Jiang, Z. Single beam Cs-Ne SERF atomic magnetometer with the laser power differential method. *Opt. Express* **2022**, *30*, 16541. [CrossRef] [PubMed]
28. Wu, W.; Zhou, B.; Liu, G.; Chen, L.; Wang, J.; Fang, J. Novel nested saddle coils used in miniature atomic sensors. *AIP Adv.* **2018**, *8*, 075126. [CrossRef]
29. Wang, J.; Zhou, B.; Wu, W.; Chen, L.; Fang, J. Uniform Field Coil Design Based on the Target-Field Method in Miniature Atomic Sensors. *IEEE Sens. J.* **2019**, *19*, 2895–2901. [CrossRef]
30. Xiao, W.; Wu, T.; Peng, X.; Guo, H. Atomic spin-exchange collisions in magnetic fields. *Phys. Rev. A* **2021**, *103*, 043116. [CrossRef]
31. Shah, V.; Romalis, M.V. Spin-Exchange-Relaxation-Free Magnetometry Using Elliptically-Polarized Light. *Phys. Rev. A* **2009**, *80*, 013416. [CrossRef]
32. Yan, Y.; Lu, J.; Zhang, S.; Lu, F.; Yin, K.; Wang, K.; Zhou, B.; Liu, G. Three-axis closed-loop optically pumped magnetometer operated in the SERF regime. *Opt. Express* **2022**, *30*, 18300. [CrossRef] [PubMed]

photonics

MDPI

Article

Precise Determination of Magnetic Gradient Relaxation of Coupled Atomic Spin Ensemble in Spin-Exchange Relaxation-Free Co-Magnetometer

Xiujie Fang [1,2], Kai Wei [1,2], Wenfeng Fan [1,2], Siran Li [1,2,3,*], Qian Cao [1,2,*], Wei Quan [1,2], Yueyang Zhai [1,2] and Zhisong Xiao [4]

1 Key Laboratory of Ultra-Weak Magnetic Field Measurement Technology, Ministry of Education, School of Instrumentation and Optoelectronic Engineering, Beihang University, Beijing 100191, China

2 Zhejiang Provincial Key Laboratory of Ultra-Weak Magnetic-Field Space and Applied Technology, Hangzhou Innovation Institute, Beihang University, Hangzhou 310051, China

3 The Shenyuan Honors College, Beihang University, Beijing 100191, China

4 School of Physics, Beihang University, Beijing 100191, China

* Correspondence: lsr2014@buaa.edu.cn (S.L.); buaacq@buaa.edu.cn (Q.C.)

Abstract: Inside a spin-exchange relaxation-free (SERF) co-magnetometer with a high-pressure buffer gas atomic cell, the magnetic field gradient causes the decoherence of atomic spins to produce magnetic-field gradient relaxation. This paper presents a new method for the accurate measurement of magnetic field gradient relaxation of alkali metal atoms and inert atoms of strongly coupled spin systems under triaxial magnetic field gradients in the K-Rb-^{21}Ne co-magnetometer. The magnetic field gradient relaxation of alkali metal atoms is measured using a step magnetic field modulation method, and the magnetic field gradient relaxation of inert atoms is measured using a combined free induction decay and spin growth method. The method does not require the use of large background magnetic fields and RF fields to maintain the atoms in the SERF state, does not require additional optics, and is not affected by the pumping or detecting of optical power. A kinetic model that considers a large electron-equivalent magnetic field was designed and a gradient relaxation model was developed. The quadratic coefficients of the experimentally measured gradient relaxation curves fit the theoretical model well over the range of the applied magnetic field gradients, confirming the validity of the proposed method.

Keywords: spin-exchange relaxation-free (SERF) co-magnetometer; magnetic field gradient relaxation; strongly coupled spin system

1. Introduction

The atomic spin inertial sensor has angular momentum based on the spin properties of the atom. It carries out Larmor precession in the presence of a magnetic field to achieve inertial measurement. It was developed based on the spin-exchange relaxation-free (SERF) effect proposed by the Happer team in 1973, in which the spin-exchange relaxation between atoms can be eliminated under high-temperature conditions with high-density atoms in a low magnetic environment [1,2]. The co-magnetometer can be divided into nuclear spin inertial measurements as inertial sensing carriers and ultrahigh-sensitive magnetometers as the final signal output to detect magnetic field signal probes. The ultrahigh-sensitive magnetometer has surpassed the superconducting quantum interference device by achieving a 0.16 $fT/Hz^{1/2}$ magnetic field measurement index [3,4]. The SERF co-magnetometer is an important development direction and research focus in the field of quantum precision measurements [5–8]. Current studies show that magnetic field gradient is an important factor affecting the performance improvement of co-magnetometers [9,10]. A magnetic field gradient produced by a light field, magnetic shield system, and temperature field exists inside the device. In an atomic vapor cell with a high-pressure buffer gas, the magnetic field gradient causes atomic spin decoherence to produce magnetic field gradient relaxation. Incoherent precession results in the attenuation of the total spin polarization and an increase in the resonance line width [11]. Theoretical analysis and accurate measurement of the influence of the triaxial magnetic field gradient on magnetic-field gradient relaxation of hybrid atomic-coupled spin systems is key to improving the co-magnetometer sensitivity.

Several conventional measurement methods for the relaxation of alkali atoms in the SERF regime have been calculated using the resonance line width [12]. Most of these are based on the synchronous pumping technique [13] and electron paramagnetic resonance [14]. In the case of the synchronous pumping method, the original optical path of the device must be changed, additional optical modulators such as choppers are invoked to switch the pumping optical field, and relaxation parameters are measured according to the transient atomic response [15,16]. This method increases the complexity of the system and introduces additional noise. Additional application of a transverse RF field or a large background field is required using the electron paramagnetic resonance method. This is contrary to the condition that atoms in the SERF regime should be in an extremely weak magnetic environment. To measure the triaxial magnetic field gradient relaxation of pure alkali metal atoms, we proposed a fast spin-exchange interaction method that uses magnetic field excitation and amplitude spectrum analysis [17]. However, for the SERF co-magnetometer, the atomic source includes alkali metal atoms and noble atoms with long relaxation times. The alkali metal atoms are pumped-polarized and subsequently transfer the angular momentum to the inert atoms through spin-exchange collisions. This results in the hyperpolarization of the inert atoms, which are strongly coupled with the alkali metal atoms, and the above-mentioned atomic magnetic field gradient relaxation measurement method is no longer applicable.

Methods for measuring the relaxation of inert atoms include spin-flip, spin-growth [18], and free-induction decay (FID) methods [19]. The spin-flip method requires the reversal of polarization by using π pulses. This method requires additional optical elements to be introduced into the device, which increases the complexity of the experiment and reduces the measurement accuracy. The signal intensity of the state in which the unpolarized status of the coupled atomic system is synthesized and gradually reaches stability was measured using the spin-growth method. The free induction decay method was used to obtain the spin-decay curve by applying a magnetic field step modulation signal to record the signal amplitude. Because applying magnetic field modulation will affect the polarization of inert atoms, the measurement accuracy is not high using the FID method alone in the experiment. The above two methods are not affected by the pumped optical power and detected optical power and do not require the addition of external measurement equipment.

The objective of this study was to design an accurate measurement method for the magnetic gradient relaxation of alkali metal and inert atoms in the strongly coupled spin-system synthesis of the SERF co-magnetometer under triaxial gradient magnetic field conditions. The magnetic field gradient relaxation of alkali metal atoms is measured using a step magnetic field modulation method, and the magnetic field gradient relaxation of inert atoms is measured using a combined free induction decay and spin growth method. The method provides a measurement means for reducing the relaxation of the magnetic field gradient in coupled atomic systems and provides data support for enhancing the self-compensating capability of coupled atomic systems and further improving the sensitivity of co-magnetometers.

2. Basic Principle

2.1. Fundamentals of Co-Magnetometer

The atomic spin system in the K-Rb-^{21}Ne co-magnetometer includes light-pumping alkali metal electron spin processes and alkali metal electron spins hyperpolarizing inert nucleus spins through spin-exchange collisions. In co-magnetometers, the polarized noble gas atoms are the key working substance for the measurements. The start-up time of the device is determined by the hyperpolarization time of the nuclear spins, and the degree of spin polarization of the inert atomic nuclei determines the dynamic range of the inertial measurements [8]. Electron spins of alkali metal atoms transfer photon energy to inert atomic nuclear spins through spin-exchange collisions. The inert nucleon polarization is low when the atomic density of the alkali metal is small, and increasing the atomic density of the alkali metal can further increase the inert nucleon polarization. When only a single alkali metal atom is present inside the gas chamber, the greater the atomic density and the greater the optical depth inside the chamber, the faster the optical decay of the pumped light, which not only reduces the efficiency of the hyperpolarized inert atoms but also brings about a large polarization gradient which affects the inertial measurement sensitivity. This problem is solved using a hybrid pumping technique, whereby two alkali metal atoms are charged into the gas chamber, one of which has a lower density and can be uniformly polarized by the pumping laser. The polarized low-density alkali metal atoms transfer their atomic spin uniformly to the high-density alkali metal via atomic spin-exchange collisions, and the polarized high-density alkali metal atoms then hyperpolarize the inert atoms by the spin-exchange optical pumping process [20].

For alkali metal atoms, the spin-collision cross-sectional area of the K atom is smaller than that of the Rb atom, but at the same experimental temperature, the saturated vapor temperature pressure density of the Rb atom is greater than that of the K atom, and the Rb atom reaches a higher density at lower experimental temperatures. The laser pumping of low-density K atoms and the polarization of high-density Rb atoms by spin-exchange collisions are therefore used for hybrid pumping [21].

Polarized alkali metal atomic spin ensembles have optical rotation properties. After the linearly polarized detection light synthesized by left-circularly polarized light and right-circularly polarized light, which is detuned to the resonance frequency, passes through the polarized atom vapor due to the birefringence effect, the polarization direction of the detected light changes. The angle of polarization change of the detected light incident along the x direction is proportional to the projection of alkali metal polarizability in the x direction. It is used to detect the magnetic field and rotation signal of the input device. The angle of rotation when the detection light is detected using the D1 line is [22]:

$$\theta = -\frac{\pi}{2} n_{Rb} l r_e c P_e f_{\mathrm{D1}} \frac{v_{\mathrm{pr}} - v_{\mathrm{D1}}}{\left(v_{\mathrm{pr}} - v_{\mathrm{D1}}\right)^2 - \left(\Gamma_{\mathrm{D1}}/2\right)^2}, \tag{1}$$

where l is the interaction distance between the probe light and the alkali metal atom, n_{Rb} is the density number of the rubidium atom, c is the propagation speed of the light, r_e is the radius of the electron. $\Gamma_{\mathrm{D1}}, f_{\mathrm{D1}}$, and v_{D1} are the pressure width, oscillation intensity, and resonance frequency of the probe light on the Rb D1 line, respectively. v_{pr} is the frequency of the probe light.

In order to suppress low-frequency noise, a photoelastic modulator (PEM) is used to detect the light in the high-frequency band to improve the detection sensitivity. The PEM's modulation amplitude is α_m, and modulation frequency is ω_m. The amount of phase delay is $\delta(t) = \alpha_m \sin(\omega_m t)$. According to Marius' law and the principle of polarized light propagation, the light intensity received by the photodetector I_t is:

$$I_t = \left[\frac{I_0 \alpha_m^2}{8} + I_0 \theta \alpha_m \sin(\omega_m t) - \frac{I_0 \alpha_m^2}{8} \cos(2\omega_m t)\right], \tag{2}$$

I_0 is the initial light intensity into the vapor cell through the deflector. Finally, PD is used to convert the optical signal into an electrical signal. The final output V_{out} of the co-magnetometer is as follows:

$$V_{out} = \eta M_{ac} I_0 e^{-oD(v)} \alpha_m \theta, \tag{3}$$

where η is the conversion factor of PD, M_{ac} is preamplifier gain, and OD is the optical depth.

2.2. Dynamics Modeling of Spin Ensembles in Magnetic Fields

The density matrix can be used to describe the mathematical model of atomic-spin system synthesis under the combined effect of the magnetic field, rotation signal, and spin coupling system in the SERF co-magnetometer. However, its solution process is too complicated. The atoms are subjected to a notably small spin-in angle by the external environment, and the Bloch equations are simplified into a linearized matrix.

In this paper, based on the SERF co-magnetometer of the K-Rb-^{21}Ne atomic source, the alkali metal atoms K and Rb are indistinguishable from each other owing to the fast atomic-spin-exchange collision process, the polarization rates of K and Rb atoms, and the signals of their responses to the magnetic field being the same. The Bloch equations of the two atoms are combined to establish the equivalent K-Rb atomic mixing equation. In this equation, when the density ratio of K and Rb atoms is $D_r = n_K/n_{Rb}$, the equivalent equation is [23,24]:

$$R_p^{Rb} = \frac{D_r R_p^K}{1+D_r}, \tag{4}$$

$$R_{se}^{ne} = \frac{D_r R_{se}^{^{21}\mathrm{Ne-K}} + R_{se}^{^{21}\mathrm{Ne-Rb}}}{1+D_r}, \tag{5}$$

$$R_{se}^{en} = R_{se}^{\mathrm{K-^{21}Ne}} + R_{se}^{\mathrm{Rb-^{21}Ne}}, \tag{6}$$

where R_p^{Rb} and R_p^K is the rate of pumping to Rb and K atoms, respectively; R_{se}^{ne}. R_{se}^{en} are the K-Rb to ^{21}Ne exchange-collision relaxation rate and ^{21}Ne to K-Rb exchange-collision relaxation rate, respectively. $R_{se}^{\mathrm{K-^{21}Ne}}$ and $R_{se}^{\mathrm{Rb-^{21}Ne}}$ are K to ^{21}Ne exchange-collision relaxation rate and Rb to ^{21}Ne exchange-collision relaxation rate, respectively.

When the pumping light is along the z-axis, and the probe light is along the x-axis, the equivalent set of Bloch equations for the mixed K-Rb atomic source with inert atom ^{21}Ne can be expressed as follows [23,25]:

$$\frac{\partial \vec{P^e}}{\partial t} = \frac{\gamma_e}{Q(P^e)}\left(\vec{B} + \lambda M_0^n \vec{P^n}\right) \times \vec{P^e} - \vec{\Omega} \times \vec{P^e} + \frac{R_p^{Rb}\vec{s_p} + R_{se}^{en}\vec{P^n}}{Q(P^e)} - \frac{\{R_1^e, R_2^e, R_2^e\}\vec{P^e}}{Q(P^e)}, \tag{7}$$

$$\frac{\partial \vec{P^n}}{\partial t} = \gamma_n\left(\vec{B} + \lambda M_0^e \vec{P^e}\right) \times \vec{P^n} - \vec{\Omega} \times \vec{P^n} + R_{se}^{ne}\vec{P^e} - \{R_1^n, R_2^n, R_2^n\}\vec{P^n}, \tag{8}$$

where P^n and P^e represent the electron and nuclear spin polarization of alkali metal and inert atoms, respectively; γ_n and γ_e represent the gyromagnetic ratio of the alkali metal and inert atoms, respectively; $Q(P^e)$ is the slowing factor for alkali metal atoms in the SERF regime; and B and Ω, respectively, denote the externally applied magnetic field and the angular velocity of rotation information. In addition, the direction of angular momentum transfer of the pumped photon is s_p; and R_1 and R_2 are the respective longitudinal and transverse relaxation rates of the atoms. Moreover, the equivalent magnetic fields engendered by the strongly joined coupling of alkali metal atoms and noble gas nuclei due to the interatomic Fermi contact interaction are $\lambda M_0^e P^e$ and $\lambda M_0^n P^n$, respectively. Equations M_0^e and M_0^n are the fully polarized magnetic moments of the alkali metal and inert atoms, respectively; λ is the enhancement factor for the Fermi contact interaction.

In the experimentally mixed atomic gas chamber, the density of inert atoms is much greater than that of the alkali metals, and the magnetic field generated by inert atoms is greater than the electron magnetic field. In a spherical atomic vapor cell, the enhancement factor for the Fermi contact interaction can be expressed as $\lambda = 2\kappa_0/3$, where κ_0 is the contact constant for the Fermi interaction between atoms. The κ_0 values for different atomic sources are listed in Table 1.

Table 1. Contact constants of Fermi interactions between different atomic sources.

Atomic Source Type	Fermi Contact Constant
K-^{21}Ne	30.8 ± 2.7 [26]
K-^{4}He	5.99 ± 0.11 [27]
Rb-^{21}Ne	32.0 ± 2.9 [28]
Rb-^{4}He	6.39 ± 0.02 [28]
Rb-^{129}Xe	644 ± 269 [29]
Cs-^{129}Xe	653 ± 20 [30]

In our experiments using the large Fermi contact constant of Rb-^{21}Ne, the equivalent magnetic field generated by the alkali metal atoms cannot be neglected during the solution of the Bloch equation. When the transverse square-wave step magnetic fields ΔB_y and ΔB_x are input, the transverse polarization of the alkali metal atoms changes ΔP_x^e according to the steady-state solution of Equation (7) as

$$\frac{\Delta P_x^e}{\Delta B_y} = \frac{(\lambda M_0^n P_z^n - \delta B_z)\delta B_z P_z^e R_2^e \gamma_e}{(\lambda M_0^n P_z^n - \delta B_z)^2 R_2^{e2} + (\lambda M_0^e P_z^e + \lambda M_0^n P_z^n - \delta B_z)^2 \delta B_z{}^2 \gamma_e{}^2}, \tag{9}$$

$$\frac{\Delta P_x^e}{\Delta B_x} = \frac{(\lambda M_0^e P_z^e + \lambda M_0^n P_z^n - \delta B_z)\delta B_z{}^2 P_z^e \gamma_e{}^2}{(\lambda M_0^n P_z^n - \delta B_z)^2 R_2^{e2} + (\lambda M_0^e P_z^e + \lambda M_0^n P_z^n - \delta B_z)^2 \delta B_z{}^2 \gamma_e{}^2}. \tag{10}$$

In the above solution, δB_z is the residual magnetic field in the z-direction, and the expression is $\delta B_z = B_z - B_z^c$; B_z is the magnetic field applied in the z-direction; B_z^c is the working self-compensating point of the SERF co-magnetometer with the expression $B_z^c = \lambda M_0^e P_z^e + \lambda M_0^n P_z^n$. At the operating point B_z^c, the strongly coupled atomic spin system has a self-compensating feature that automatically compensates for the disturbing magnetic field noise introduced by the external environment. It has the best dynamic performance currently [31].

Figure 1 shows the output response curves of the device obtained using the transverse square-wave modulation method for different residual magnetic fields, δB_z, under simulated conditions.

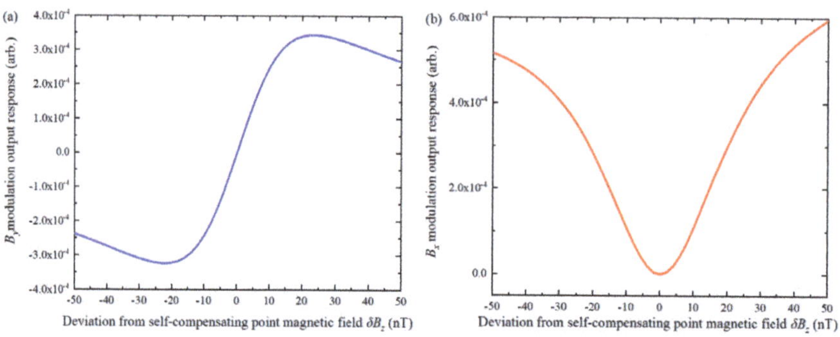

Figure 1. B_y and B_x step magnetic field modulation output response of SERF co-magnetometer. (**a**) B_y modulation output response at different residual magnetic fields. (**b**) B_x modulation output response at different residual magnetic fields.

By changing the residual magnetic field, δB_z, when applying B_x square-wave magnetic field modulation, the output response of the inertial device is an absorption curve line, and the output response is a Lorentzian line when modulated by a B_y square-wave magnetic field. The transverse relaxation parameters of the alkali metal atoms R_2^e in the strongly coupled atomic source can be obtained using Equations (9) and (10).

2.3. Magnetic Gradient Relaxation Theory for Coupled Atomic Spin Systems

Here, the alkali metal atomic relaxation rate and inert atomic relaxation rate in the SERF co-magnetometer system are discussed. For alkali metals, the atomic relaxation rates are divided into transverse and longitudinal relaxation rates, where the longitudinal relaxation rates of alkali metal atoms are composed as follows [32]:

$$R_1^e = \frac{1}{Q(P^e)}\left(R_{\mathrm{sd}}^e + R_{\mathrm{p}} + R_{\mathrm{pr}}\right) + R_{\mathrm{wall}}^e + R_{1\Delta\mathrm{B}}^e, \tag{11}$$

where R_{sd}^e is the spin-destroying collisional relaxation, which occurs when there are collisions between alkali metal atoms, between alkali metal atoms and inert atoms, and between alkali metal atoms and buffer and quenching gases. R_{wall}^e is the collisional relaxation of alkali metal atoms colliding with the walls of the gas chamber, where the collision causes the atoms to incur decoherence and to relax rapidly. $R_{1\Delta\mathrm{B}}^e$ is a longitudinal gradient relaxation of alkali metal atoms.

For the lateral relaxation of alkali metal atoms, the factors causing atomic spin precession decoherence act on the lateral relaxation, which is expressed as

$$R_2^e = R_1^e + R_{\mathrm{SE}} + R_{2\Delta\mathrm{B}}^e, \tag{12}$$

where R_{SE} is the spin-exchange collisional relaxation between atoms. $R_{2\Delta\mathrm{B}}^e$ is a transverse gradient relaxation of alkali metal atoms.

When the alkali metal atomic density is high and, in a notably low magnetic field experimental environment, the spin-exchange collisional relaxation of alkali metal atoms can be suppressed or even eliminated when the spin-exchange collisional rate is much higher than the spin-Larmor feeding frequency of the alkali metal atoms in the SERF state. Alkali metal atoms in the SERF operating state exhibit a spin-temperature distribution with a spin-exchange relaxation rate expressed as [33]

$$R_{\mathrm{SE}} = \frac{{\gamma_e}^2 B^2 (Q(P^e)^2 - (2I+1)^2)}{2Q(P^e)^2 R_{se}^e}. \tag{13}$$

Inside the inertial measurement device, the pumped optical field polarization rate gradient, the magnetic shielding static magnetic field gradient, and the residual magnetic field gradient of the heating system result in the decoherence of the atomic spins. The large value of the main magnetic field, the alkali metal atomic spin longitudinal magnetic field gradient relaxation, and the transverse magnetic field gradient relaxation are expressed as [17]

$$R_{1\Delta\mathrm{B}}^e = 2D\frac{\left|\vec{\nabla} B_x\right|^2 + \left|\vec{\nabla} B_y\right|^2}{B_0^2}, \tag{14}$$

$$R_{2\Delta\mathrm{B}}^e = \frac{8{\gamma^e}^2 r^4 \left|\vec{\nabla} B_z\right|^2}{175DQ(P_z)^2}, \tag{15}$$

where B_0 is the value of the main magnetic field incurred by atoms on the z-axis, r is the radius of the atomic vapor cell, and D is the temperature- and pressure-dependent diffusion coefficients.

The ^{21}Ne noble atomic relaxation rate in the K-Rb-^{21}Ne co-magnetometer is

$$R_{rel}^n = R_{\mathrm{se}}^{en} + R_{\mathrm{quad}}^n + R_{\mathrm{sd}}^n + R_{\Delta\mathrm{B}}^n, \tag{16}$$

where R_{quad}^n is the electric quadrupole moment relaxation resulting from collisions between ^{21}Ne atoms, which is related to the atomic density. $R_{\Delta\mathrm{B}}^n$ is the magnetic gradient relaxation term of ^{21}Ne in the magnetic field gradient, where the longitudinal magnetic field gradient relaxation term is like those of the alkali metal atoms, and the transverse magnetic field gradient relaxation is the following [34,35]:

$$R^n_{1\Delta B} = 2D\frac{\left|\vec{\nabla} B_x\right|^2 + \left|\vec{\nabla} B_y\right|^2}{B_0^2}, \tag{17}$$

$$R^n_{2\Delta B} = \frac{8\gamma^{e2}r^4\left|\vec{\nabla} B_z\right|^2}{175D}, \tag{18}$$

By means of the above-mentioned modeling of the dynamics of the strongly coupled atomic spin system and the relaxation model under magnetic field gradients, the magnetic gradient relaxation of alkali metal atoms and inert atoms was accurately measured. More specifically, a triaxial magnetic field gradient was applied through a gradient coil, and a step magnetic field modulation method and spin-growth method combined with the FID method were applied.

3. Experimental Setup

A diagram of the experimental setup is shown in Figure 2. The device is divided into four main subsystems: a sensitive head system, an electronic measurement and heating system, magnetic shielding and magnetic compensation system, and a pumping and probe light system. The sensitive head system is at the center of the unit and includes a mixed atomic vapor cell, a non-magnetic vacuum structure, and a water-cooled structure. A spherical vapor cell with a 6 mm radius contains K and Rb with a density ratio Dr = 1:86. The mixed alkali metal atomic density is $n_e = 7.51 \times 10^{13}$ cm^{-3}; 15 Torr N_2 is the quench gas, and 2080Torr ^{21}Ne. The atomic vapor cell is made of aluminosilicate glass. Non-magnetic materials are used inside the device to keep the atoms in extremely weak magnetic field experimental conditions, ensuring that the atoms are always in the SERF state. The non-magnetic vacuum structure is mainly used to ensure temperature stability so that the atomic parameters in the atomic gas chamber are fixed values and prevent uneven distribution of atomic density in the atomic gas chamber caused by temperature gradient generated by heating. The water-cooled part is wound around the outside of the vacuum cavity to reduce temperature, prevent the high temperature from affecting magnetic shielding and magnetic compensation performance, and avoid introducing new thermal magnetic noise.

The cells were placed in a ceramic oven controlled by a proportional-integral conductor at 80 kHz AC and heated to 190 °C. At the same time, in order to avoid generating additional magnetic field noise, the heating line is twisted to cancel the magnetic field of the heating coil. The cell temperature was monitored using a PT1000 resistor with a temperature control accuracy of $\pm$5 mK. Insulation outside the oven was achieved with a non-magnetic polyether ether ketone vacuum cavity. The device required an extremely weak magnetic field condition—ensuring that the atoms were in a SERF state—using a magnetic shielding system with low magnetic noise and a high magnetic shielding factor [36,37]. The magnetic shielding system consisted of an outer layer of four μ-metals with a shielding factor of 10^5, as well as a residual field of 0.5 nT and a magnetic noise of 5 fT/Hz$^{1/2}$ inside the device after shielding [38]. A set of triaxial compensation coils with a set of triaxial gradient coils driven by a function generator (33500B, Keysight, Santa Rosa, CA, USA) was used.

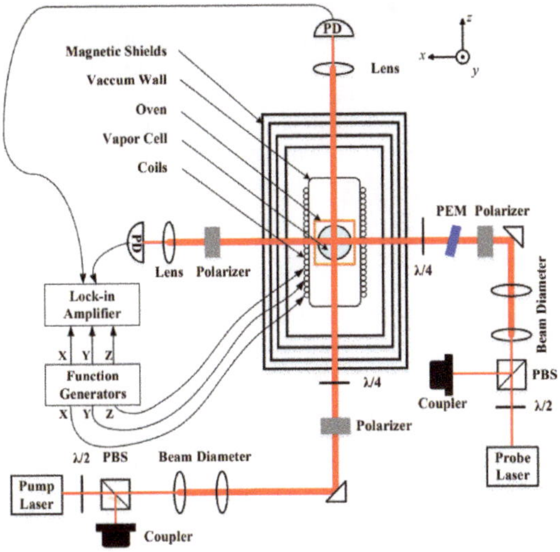

Figure 2. Experimental setup. λ/4: quarter-wave plate. λ/2: half-wave plate. PD: photodetector. PBS: polarized beam splitter. PEM: photoelastic modulator.

Both the pumping and probe lights were generated by the distributed Bragg reflector laser. The pumping laser was beaming split by a PBS, and one of the beams was used to monitor the laser wavelength. The beam radius was expanded to 7 mm by a lens beam spreading system and then passed through a quarter-wave to circularly polarize light for pumping the alkali metal atoms. The pumping power was 412 mW, and the wavelength was locked to the D1 line of the K atom at 770.107 nm. The probe light beam radius was 2 mm, the power was 11.3 mW, and the wavelength was 795.511 nm. It was detuned away from the D1 line of Rb atoms. To isolate low-frequency noise and improve the probe sensitivity, a photoelastic modulator (PEM) probe method was used. The PEM model used in the experiments was from I/FS Hinds Instruments with a modulation frequency of 50 kHz and a modulation angle of 0.08 rad. The co-magnetometer mainly focuses on low-frequency signals, so it needs to use a low-frequency non-magnetic vibration isolation device to isolate the vibration in the environment so that the vibration noise felt by the optical platform on the isolator is as small as possible.

Figure 2 shows the pumping light along the z-axis and the probe light along the x-axis. Polarized atomic spin ensembles exhibit optical rotation properties. After the linearly polarized probe light passes through the polarized atomic cell, the polarization direction of the probe light changes owing to the birefringence effect [39], where the refractive indices of the left- and right-handed polarized light are different. The angle of change in the polarization of the detection light incident along the x-direction is proportional to the projection of the alkali metal polarization rate in the x-direction for detecting the input device signal. The light rotation angle information is measured by a photodetector; the signal is amplified by a lock-in amplifier (HF2LI, Zurich Instruments, Zurich, Switzerland), and the optical signal is converted into a voltage signal. The voltage signal is then collected and processed using DAQ.

4. Results and Discussion

4.1. Magnetic Gradient Relaxation of Alkali Metal Atom Measurement

The residual magnetic field in the device was compensated with a DC magnetic field using triaxial compensation coils to identify the self-compensating operating point, where $\delta B_z = 0$. The magnetic gradient relaxation of alkali metal atoms in strongly coupled ensembles was measured. When no gradient magnetic field was applied, a step magnetic field modulation was applied along the y-axis, the magnitude of the residual bias magnetic field δB_z value in the z-axis was changed, and the difference in the output signal response amplitude was measured. Accordingly, the transverse relaxation rate of the alkali metal atoms was measured according to Equation (9). From Equations (14) and (15), the transverse gradient magnetic field was related to the longitudinal magnetic gradient relaxation, and the longitudinal gradient magnetic field was related to the transverse magnetic gradient relaxation when the triaxial gradient magnetic field was applied separately. Equation (12) can be expressed as follows:

$$R_2^e = R_{1-0}^e + R_{SE} + R_{1\Delta B}^e + R_{2\Delta B}^e, \tag{19}$$

where R_{1-0}^e is the fixed relaxation value other than the spin-exchange relaxation rate, R_{SE} is the spin-exchange collisional relaxation between atoms before the triaxial magnetic field gradient is applied to the device. When no magnetic field gradient is applied, $R_{1\Delta B}^e = 0$, $R_{2\Delta B}^e = 0$.

The value of R_2^e can be obtained from the description in Section 2.2 by changing the residual magnetic field when applying square-wave magnetic field modulation. The slowing factor $Q(P^e)$ could be obtained by measuring the polarizability of alkali metal atoms in coupled spin ensembles based on the steady-state AC response [40], and the value of R_{SE} was calculated according to Equation (13). A fixed value of $R_{1-0}^e = R_2^e - R_{SE}$ was obtained according to the above steps. For the experimental measurements without magnetic field gradients, the transverse relaxation of the alkali metal atoms was at the self-compensating point $R_2^e = 4020.67\ s^{-1}$, the theoretical spin-exchange relaxation rate was $R_{SE} = 189.57\ s^{-1}$ with a fixed relaxation value $R_{1-0}^e = 3831.43\ s^{-1}$. The value of R_{1-0}^e was maintained constant in subsequent experiments with magnetic field gradients.

Subsequently, a triaxial magnetic field gradient ranging from 5 to 25 nT/cm was applied separately. The above steps were repeated to obtain an accurate measurement of the magnetic gradient relaxation of alkali metal atoms by measuring the transverse relaxation rate, measuring the slowing down factor, and calculating the spin-exchange relaxation rate. Figure 3 shows the difference between the highest and lowest output voltage of the device signal corresponding to the different residual magnetic field points of the step magnetic field modulation in the case of an applied magnetic field gradient along the z-axis and step-modulated magnetic field with amplitude 0.5 nT and frequency 0.2 Hz.

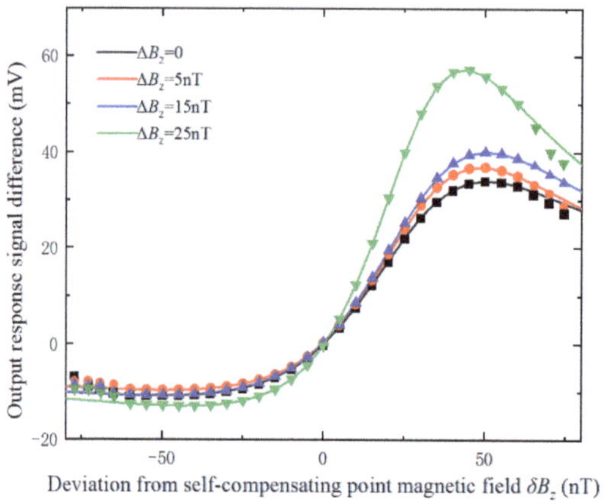

Figure 3. Experimental measurements of the stepped magnetic field modulated output response with an applied z-axis magnetic field gradient.

The results of the experimental measurements of the relaxation of the magnetic-field gradients of the alkali metal atoms in the applied x- and y-axis magnetic-field gradients are shown in Tables 2 and 3.

Table 2. Experimental results of magnetic gradient relaxation of alkali metal atoms in coupled spin ensemble under applied *x*-axis magnetic gradient.

Magnetic Field Gradient	5 nT/cm	10 nT/cm	15 nT/cm	20 nT/cm	25 nT/cm
R_2^e	4032.65 s^{-1}	4048.53 s^{-1}	4066.20 s^{-1}	4085.12 s^{-1}	4103.62 s^{-1}
R_{SE}	192.81 s^{-1}	208.91 s^{-1}	226.51 s^{-1}	244.77 s^{-1}	264.3 s^{-1}
$R^e_{1\Delta B_x}$	8.41 s^{-1}	8.13 s^{-1}	8.26 s^{-1}	8.92 s^{-1}	7.89 s^{-1}

Table 3. Experimental results of magnetic gradient relaxation of alkali metal atoms in coupled spin ensemble under applied *y*-axis magnetic gradient.

Magnetic Field Gradient	5 nT/cm	10 nT/cm	15 nT/cm	20 nT/cm	25 nT/cm
R_2^e	4051.22 s^{-1}	4069.02 s^{-1}	4072.18 s^{-1}	4089.65 s^{-1}	4094.43 s^{-1}
R_{SE}	160.57 s^{-1}	175.06 s^{-1}	183.20 s^{-1}	193.51 s^{-1}	204.96 s^{-1}
$R^e_{1\Delta B_x}$	59.22 s^{-1}	62.53 s^{-1}	57.55 s^{-1}	62.71 s^{-1}	58.04 s^{-1}

As observed in the above table, when only the transverse magnetic field gradient is applied, the relaxation of the magnetic field gradient of the alkali metal atoms does not change over the range of the applied magnetic field gradient. In the SERF co-magnetometer, both the alkali metal atoms and inert atoms experience the equivalent magnetic field generated by the atoms themselves in the self-compensating operating state owing to Fermi interactions for large main magnetic field conditions. According to Equation (14), in the range of 5 nT/cm to 25 nT/cm of transverse magnetic field gradients, the relaxation time of the magnetic-field gradient of the alkali metal atoms is on the order of milliseconds for this relaxation term to be undetectable.

The experimental measurements of the magnetic gradient relaxation with the *z*-axis magnetic field gradient applied alone are shown in Figure 4. The measured value is larger than the theoretical value owing to an increase in the longitudinal magnetic field gradient value along a quadratic curve. Here, the theoretical value quadratic coefficient is 0.0459, the experimental value quadratic coefficient is 0.0444, and the experimental measurement gradient relaxation changes in line with the theory, proving the correctness and validity of the proposed method.

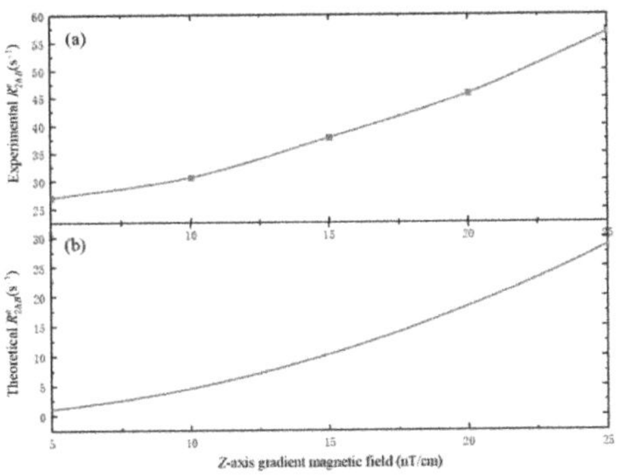

Figure 4. Experimental results of magnetic gradient relaxation of alkali metal atoms in coupled spin ensemble under applied *z*-axis magnetic gradient. (**a**) The experimental measurement results; (**b**) The simulation results based on the theoretical Equation (15).

4.2. Magnetic Gradient Relaxation of Inert Atom Measurement

The experimental measurements of the relaxation of the magnetic field gradient of inert atoms in a strongly coupled atomic system are presented below. The Bloch equation for inert atoms shows that when a transverse step magnetic field is applied, the solution to Equation (8) is:

$$P_x^n = P_z^n e^{-tR_2^n} \sin(\gamma^n B_0 t), \tag{20}$$

$$P_z^n = P_{z0}^n \left[1 - e^{-tR_1^n} \left(1 - \frac{P_{z0}^n}{P_0^n}\right)\right], \tag{21}$$

where t is the measurement time and P_0^n is the initial polarization rate of the inert atom is:

$$P_{z0}^n = \frac{R_{se}^{en}}{R_{se}^{en} + R_{rel}^n} P^e, \tag{22}$$

The above equation facilitates the measurement of the transverse and longitudinal relaxation rates of inert atoms using the FID method combined with the spin growth method. The inert atoms are coupled to the alkali metal atoms by hyperfine interactions, and the polarized noble gas atoms generate an equivalent magnetic field acting on the alkali metal atoms so that the incoming ^{21}Ne atoms can be detected by the in-situ atomic magnetometer composed of Rb atoms, which causes a change in the output signal of the co-magnetometer. The main magnetic field in the experiment is the value of the magnetic field at the self-compensating operating point, the step magnetic field is applied along the y-axis, and the device output signal and output amplitude fitting equations are:

$$S(t) = A \sin(\omega_0 t) e^{-tR_2^n} + S_0, \tag{23}$$

$$A(t_{pump}) = k\left(1 - e^{-t_{pump} R_1^n}\right), \tag{24}$$

where S_0 is the output signal bias, t_{pump} is the optical pumping time, and k is the scaling factor related to the initial polarization rate of the inert atoms. Figure 5 shows the experimental measurement of R_2^n of the inert atom without an applied magnetic field gradient. The experimental measurements fit the theoretical equation curve well, with a fit factor greater than 0.994, and the transverse relaxation of the inert atom without the applied magnetic field gradient is 0.065 s^{-1}.

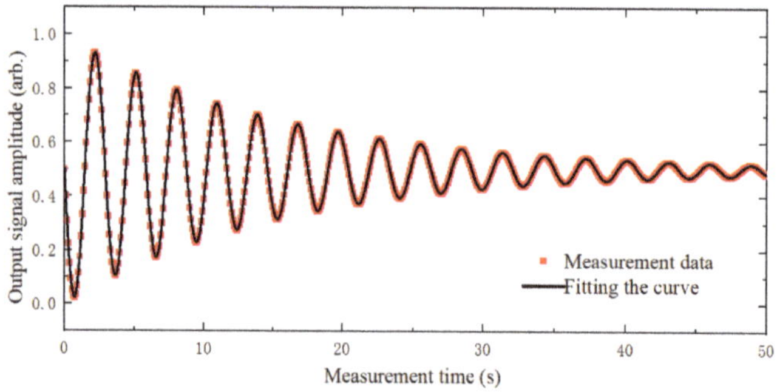

Figure 5. Experimental measurement of transverse relaxation rate of inert atoms in strongly coupled spin ensemble without magnetic field gradient.

This method, combined with the spin growth method, enables both the measurement of the transverse relaxation rate of inert atoms and the measurement of the R_1^n of inert atoms. According to Equations (20) and (21), the output signal amplitude in the measured FID curve is related to the polarization rate and the initial polarization rate of inert atoms, and

R_1^n of inert atoms is obtained from the relationship between the output signal amplitude and the optical pumping time.

The experimental procedure involved recording the amplitude values at regular intervals after the pumping light was switched on and until the polarization of the inert nuclei was completed. Figure 6 shows the results of the R_1^n measurements of inert atoms in the inertial atomic system. R_1^n of the inert atoms without application of the gradient coil is 0.00063 s^{-1}, which is a larger value than the R_{quad}^n from the collisions between the ^{21}Ne atoms calculated using the theoretical equation. This finding indicates the existence of gradient relaxation due to the transverse magnetic field gradient inside the device.

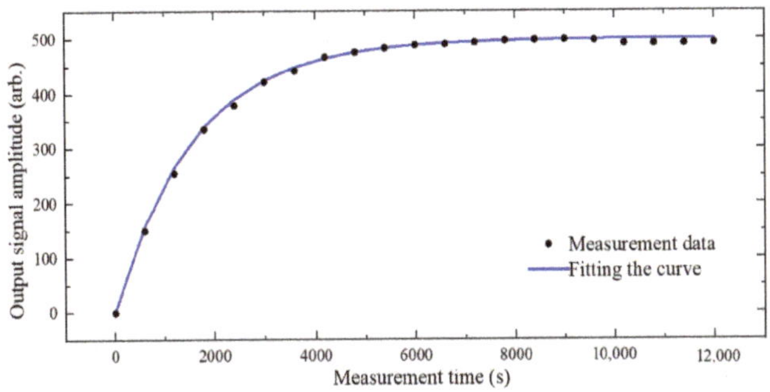

Figure 6. Experimental measurement of longitudinal relaxation rate of inert atoms in strongly coupled spin ensemble without magnetic field gradient.

The magnetic field gradient relaxation could then be accurately measured by applying a triaxial magnetic field gradient separately in the device and measuring the change in the transverse and longitudinal relaxation rates of the inert atoms using the method described above. Table 4 shows the measured R_2^n and R_1^n when the *x*-, *y*-, and *z*-axis magnetic field gradients were applied separately.

Table 4. Experimental results of R_2^n and R_1^n when the *x*-, *y*-, and *z*- axis magnetic field gradients are applied separately.

Magnetic Field Gradient (*x*-axis)	5 nT/cm	10 nT/cm	15 nT/cm	20 nT/cm	25 nT/cm
R_2^n	0.0660 s^{-1}	0.0678 s^{-1}	0.0650 s^{-1}	0.0661 s^{-1}	0.0659 s^{-1}
R_1^n	0.000826 s^{-1}	0.00104 s^{-1}	0.00128 s^{-1}	0.000158 s^{-1}	0.000189 s^{-1}
Magnetic Field Gradient (*y*-axis)	**5 nT/cm**	**10 nT/cm**	**15 nT/cm**	**20 nT/cm**	**25 nT/cm**
R_2^n	0.0682 s^{-1}	0.0686 s^{-1}	0.070 s^{-1}	0.0682 s^{-1}	0.0679 s^{-1}
R_1^n	0.000864 s^{-1}	0.00107 s^{-1}	0.00133 s^{-1}	0.000159 s^{-1}	0.000196 s^{-1}
Magnetic Field Gradient (*z*-axis)	**5 nT/cm**	**10 nT/cm**	**15 nT/cm**	**20 nT/cm**	**25 nT/cm**
R_2^n	0.128 s^{-1}	0.151 s^{-1}	0.179 s^{-1}	0.208 s^{-1}	0.242 s^{-1}
R_1^n	0.000639 s^{-1}	0.000641 s^{-1}	0.000648 s^{-1}	0.000657 s^{-1}	0.000638 s^{-1}

It is clear from the measurements that R_2^n is almost constant when the *x*-axis and *y*-axis magnetic field gradients are applied separately, and that R_1^n is constant when the *z*-axis magnetic field gradient is applied alone.

Section 4.2 and Figure 8 show the experimental measurements of the longitudinal magnetic gradient relaxation time of inert atoms $R_{1\Delta B}^n$ when the *x*- and *y*-axis magnetic field gradients are applied separately, the transverse magnetic gradient relaxation time of inert atoms $R^n_{2\Delta B}$ when the z-axis magnetic field gradients are applied alone, and the curves calculated by the theoretical formula.

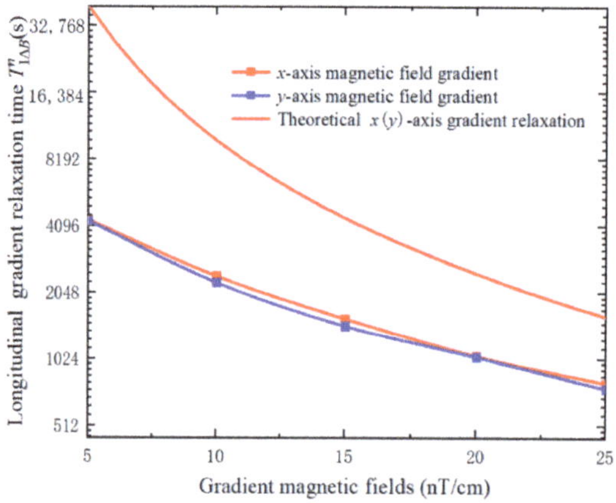

Figure 7. Measured relaxation times of the longitudinal magnetic gradient of inert atoms with an applied x- and y-axis magnetic field gradient, respectively, and curves calculated by the theoretical equations.

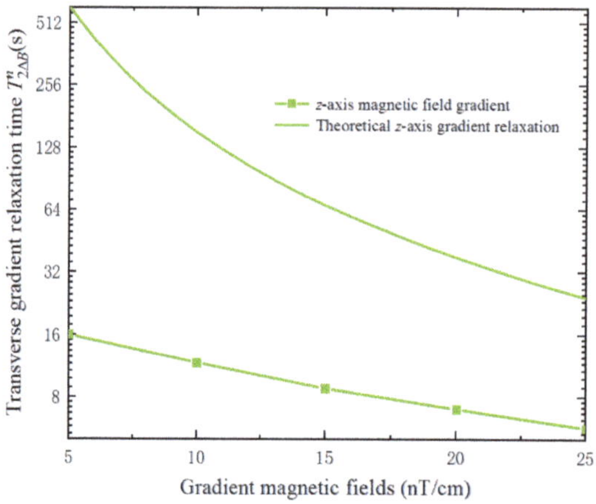

Figure 8. Measured relaxation times of transverse magnetic field gradients of inert atoms with z-axis magnetic field gradients applied alone and curves calculated from theoretical equations.

From the above experimental results, the quadratic coefficients of the longitudinal and transverse magnetic field gradient relaxation rate variation curves were 1×10^{-6} and 7×10^{-5}, respectively, and the variation patterns and magnitudes were the same as those of the theoretical model. The experimental measurement of the relaxation time was considerably lower than the value calculated by the theoretical equation. The theoretical values are calculated using a triaxial gradient magnetic field applied in the range of 5 nT/cm to 25 nT/cm. Before the application of the gradient coil, the inherent magnetic field gradient in the x- and y-axes inside the co-magnetometer was 5–10 nT/cm, and the inherent magnetic field gradient in the z-axis of the device was 20–25 nT/cm. In addition, in the measurement environment, the electrical heating system, the coupling of the coils, and the mutual coupling between the coils and the magnetic shield can generate other order gradients of magnetic fields. Improving the linearity of the compensation and magnetic field gradient coils can further improve the accuracy of the proposed measurement method.

5. Conclusions

In this paper, precise measurements and analyses of the magnetic gradient relaxation of strongly coupled spin-system-synthesized alkali metal atoms and inert atoms in the SERF co-magnetometer under a triaxial magnetic field gradient were presented. In this study, a kinetic model for the large electron equivalent magnetic field of alkali metal atoms was developed and solved. A model for the relaxation of the magnetic field gradient of a coupled atomic system was developed. The magnetic gradient relaxation of alkali metal atoms was measured using the device response to the step magnetic field modulation method, and the magnetic gradient relaxation of inert atoms was measured using a combination of FID and spin-growth methods.

The proposed method does not require the application of a large background magnetic field, and it changes the optical path with additional optics. The atoms are guaranteed to be in the SERF state, and the measurements are independent of the pumping and detection powers. The experimental quadratic coefficient of magnetic field gradient relaxation for alkali metal atoms was 0.0444, and the theoretical quadratic coefficient was 0.0459 over the range of applied magnetic field gradients. The quadratic coefficients of the longitudinal and transverse magnetic field gradient relaxation variation curves measured experimentally for inert atoms were 1×10^{-6} and 7×10^{-5}, respectively, which well aligned with the theoretical model. The validity of the proposed method was thus confirmed. The experimental results demonstrated that, in the SERF co-magnetometer, the longitudinal magnetic field gradient acts mainly on the transverse magnetic gradient relaxation of the alkali metal and inert atoms, and the transverse magnetic gradient acts mainly on the longitudinal magnetic gradient relaxation of the inert atoms. Based on the research presented herein, we suggest the design of new methods to suppress the magnetic field gradient in the SERF co-magnetometer to improve its dynamic performance and sensitivity. This will be the focus of future research.

Author Contributions: Conceptualization, X.F. and S.L.; methodology, K.W.; validation, X.F., Q.C., W.F. and S.L.; formal analysis, X.F.; investigation, X.F. and S.L.; data curation, X.F.; writing—original draft preparation, X.F.; writing—review and editing, Q.C. and W.F.; project administration, W.Q., Y.Z. and Z.X.; funding acquisition, W.Q. and W.F. All authors have read and agreed to the published version of the manuscript.

Funding: This work was supported by the National Science Fund for Distinguished Young Scholars 61925301, the National Natural Science Foundation of China under Grant No. 61903013, No. 62103026, and No. 61975005 and Outstanding Research Project of Shen Yuan Honors College, BUAA (230122102).

Institutional Review Board Statement: Not applicable.

Informed Consent Statement: Not applicable.

Data Availability Statement: The data presented in this study are available on reasonable request from the corresponding author.

Conflicts of Interest: The authors declare no conflict of interest.

References

1. Happer, W.; Tam, A.C. Effect of rapid spin exchange on the magnetic-resonance spectrum of alkali vapors. *Phys. Rev. A.* **1977**, *16*, 5. [CrossRef]
2. Meyer, D.; Larsen, M. Nuclear Magnetic Resonance Gyro for Inertial Navigation. *Gyroscopy Navig.* **2014**, *5*, 75. [CrossRef]
3. Kominis, I.K.; Kornack, T.W.; Allred, J.C.; Romalis, M.V. A Subfemtotesla Multichannel Atomic Magnetometer. *Nature* **2003**, *422*, 596. [CrossRef] [PubMed]
4. Dang, H.B.; Maloof, A.C.; Romalis, M.V. Ultrahigh Sensitivity Magnetic Field and Magnetization Measurements with an Atomic Magnetometer. *Appl. Phys. Lett.* **2010**, *97*, 151110. [CrossRef]
5. Larsen, M.; Bulatowicz, M. Nuclear Magnetic Resonance Gyroscope for DARPA's Micro-Technology for Positioning, Navigation and Timing Program. In Proceedings of the 2014 International Symposium on Inertial Sensors and Systems (ISISS), Laguna Beach, CA, USA, 25–26 February 2014; p. 1.
6. Stoner, R.; Walsworth, R. Atomic physics: Collisions give sense of direction. *Nat. Phys.* **2006**, *2*, 17. [CrossRef]

7. Fang, J.C.; Qin, J. Advances in Atomic Gyroscopes: A View from Inertial Navigation Applications. *Sensors* **2012**, *12*, 6331. [CrossRef]
8. Zhao, T.; Ying, Y.; Wei, K.; Xie, H.T.; Mu, T.J.; Fang, X.J.; Xu, Z.T.; Zhai, Y.Y. Ultra-sensitive all-optical comagnetometer with laser heating. *J. Phys. D* **2022**, *55*, 165103. [CrossRef]
9. Wei, K.; Zhao, T.; Fang, X.J.; Xu, Z.T.; Liu, C.; Cao, Q.; Wickenbrock, A.; Hu, Y.H.; Ji, W.; Fang, J.C.; et al. Ultrasensitive Atomic Comagnetometer with Enhanced Nuclear Spin Coherence. *Phys. Rev. Lett.* **2023**, *130*, 63201. [CrossRef]
10. Pang, H.Y.; Liu, F.; Fan, W.F.; Wu, Z.H.; Yuan, Q.; Quan, W. Comprehensive analysis of the effects of magnetic fieldgradient on the performance of the SERF co-magnetometer. *Opt. Express* **2023**, *31*, 5215. [CrossRef]
11. Schearer, L.D.; Walters, G.K. Nuclear Spin-Lattice Relaxation in the Presence of Magnetic-Field Gradients. *Phys. Rev. J.* **1965**, *139*, 1398. [CrossRef]
12. Happer, W.; Tang, H. Spin-exchange shift and narrowing of magnetic resonance lines in optically pumped alkali vapors. *Phys. Rev. Lett.* **1973**, *31*, 273. [CrossRef]
13. Allred, J.C.; Lyman, R.N.; Kornack, T.W.; Romalis, M.V. High-sensitivity atomic magnetometer unaffected by spin-exchange relaxation. *Phys. Rev. Lett.* **2002**, *89*, 130801. [CrossRef]
14. Ding, Z.C.; Long, X.W.; Yuan, J.; Fan, Z.F.; Luo, H. Sensitive determination of the spin polarization of optically pumped alkali-metal atoms using near-resonant light. *Sci. Rep.* **2016**, *6*, 32605. [CrossRef]
15. Li, R.J.; Quan, W.; Fang, J.C. Polarization Measurement of Cs Using the Pump Laser Beam. *IEEE Photonics J.* **2017**, *9*, 1. [CrossRef]
16. Ma, Y.N.; Zhang, K.X.; Wang, Y.G.; Yang, K.; Zhai, Y.Y.; Lu, J.X. Fast extraction of the electron spin-relaxationrate in the SERF magnetometer from a transient response. *Opt. Express* **2022**, *30*, 17383. [CrossRef]
17. Fang, X.J.; Wei, K.; Zhai, Y.Y.; Zhao, T.; Chen, X.; Zhou, M.T.; Liu, Y.; Ma, D.Y.; Xiao, Z.S. Analysis of effects of magnetic field gradient on atomic spin polarization and relaxation in optically pumped atomic magnetometers. *Opt. Express* **2022**, *30*, 3926. [CrossRef]
18. Mirijanian, J.J. Techniques to Characterize Vapor Cell Performance for a Nuclear-Magnetic-Resonance Gyroscope. Ph.D. Thesis, California Polytechnic State University, San Luis Obispo, CA, USA, 2012.
19. Savukov, I.M.; Romalis, M.V. NMR Detection with an Atomic Magnetometer. *Phys. Rev. Lett.* **2005**, *94*, 123001. [CrossRef]
20. Zhao, T.; Zhai, Y.Y.; Liu, C.; Xie, H.T.; Cao, Q.; Fang, X.J. Spin polarization characteristics of hybrid optically pumped comagnetometers with different density ratios. *Opt. Express* **2021**, *30*, 15. [CrossRef]
21. Li, Y.; Liu, X.J.; Cai, H.W.; Ding, M.; Fang, J.C. Optimization of the alkali-metal density ratio in a hybrid optical pumping atomic magnetometer. *Meas. Sci. Technol.* **2019**, *30*, 15005. [CrossRef]
22. Duan, L.H.; Fang, J.C.; Li, R.J.; Jiang, L.W.; Ding, M.; Wang, W. Light intensity stabilization based on the second harmonic of the photo elastic modulator detection in the atomic magnetometer. *Opt. Express* **2015**, *23*, 32481. [CrossRef]
23. Fang, J.C.; Chen, Y.; Lu, Y.; Quan, W.; Zou, S. Dynamics of Rb and ^{21}Ne spin ensembles interacting by spin exchange with a high Rb magnetic field. *J. Phys. B At. Mol. Opt. Phys.* **2016**, *49*, 135002. [CrossRef]
24. Chen, Y.; Quan, W.; Zou, S.; Lu, Y.; Duan, L.H.; Li, Y.; Zhang, H.; Ding, M.; Fang, J.C. Spin exchange broadening of magnetic resonance lines in a high-sensitivity rotating K-Rb-^{21}Ne comagnetometer. *Sci. Rep.* **2016**, *6*, 36547. [CrossRef] [PubMed]
25. Kornack, T.W.; Ghosh, R.K.; Romalis, M.V. Nuclear spin gyroscope based on an atomic comagnetometer. *Phys. Rev. Lett.* **2005**, *95*, 230801. [CrossRef] [PubMed]
26. Ghosh, R.K. Spin Exchange Optical Pumping of Neon and Its Applications. Ph.D. Thesis, Princeton University, Princeton, NJ, USA, 2009.
27. Babcock, E.; Nelson, I.A.; Kadlecek, S.; Walker, T.G. ^{3}He Polarization-Dependent EPR Frequency Shifts of Alkali-Metal–^{3}He Pairs. *Phys. Rev. A* **2005**, *71*, 13414. [CrossRef]
28. Stoner, R.E.; Walsworth, R.L. Measurement of the 21Ne Zeeman Frequency Shift Due to Rb-^{21}Ne Collisions. *Phys. Rev. A* **2002**, *66*, 32704. [CrossRef]
29. Schaefer, S.R.; Cates, G.D.; Chien, T.R.; Gonatas, D.; Happer, W.; Walker, T.G. Frequency Shifts of the Magnetic-Resonance Spectrum of Mixtures of Nuclear Spin-Polarized Noble Gases and Vapors of Spin-Polarized Alkali-Metal Atoms. *Phys. Rev. A* **1989**, *39*, 5613. [CrossRef]
30. Fang, J.C.; Wan, S.-A.; Chen, Y. Measurement of ^{129}Xe Frequency Shift Due to Cs-^{129}Xe Collisions. *Chin. Phys. B* **2014**, *23*, 63401. [CrossRef]
31. Wei, K.; Zhao, T.; Fang, X.J.; Li, H.R.; Zhai, Y.Y.; Han, B.C.; Quan, W. Simultaneous Determination of the Spin Polarizations of Noble-Gas and Alkali-Metal Atoms Based on the Dynamics of the Spin Ensembles. *Phys. Rev. A* **2020**, *13*, 44027. [CrossRef]
32. Seltzer, S.J. Developments in Alkali-Metal Atomic Magnetometry. Ph.D. Thesis, Princeton University, Princeton, NJ, USA, 2008.
33. Kornack, T.W.; Romalis, M.V. Dynamics of Two Overlapping Spin Ensembles Interacting by Spin Exchange. *Phys. Rev. Lett.* **2002**, *89*, 253002. [CrossRef]
34. Cates, G.D.; Schaefer, S.R.; Happer, W. Relaxation of spins due to field inhomogeneities in gaseous samples at low magnetic field and low pressures. *Phys. Rev. A* **1988**, *37*, 2877. [CrossRef]
35. Hasson, K.C.; Cates, G.D.; Lerman, K.; Bogorad, P.; Happer, W. Spin relaxation due to magnetic-field inhomogeneities: Quartic dependence and diffusion-constant measurements. *Phys. Rev. A* **1990**, *42*, 5766. [CrossRef]
36. Lu, J.X.; Zhang, S.W.; Zhou, Y.; Yan, Y.G.; Lu, F.; Wang, K.; Zhai, Y.Y.; Ye, M. Optimal buffer gas pressure in dual-beam spin-exchange relaxation-free magnetometers. *Sens. Actuator A Phys.* **2022**, *347*, 113928. [CrossRef]

37. Yan, Y.G.; Lu, J.X.; Zhang, S.W.; Lu, F.; Yin, K.F.; Wang, K.; Zhou, B.Q.; Liu, G. Three-axis closed-loop optically pumped magnetometer operated in the SERF regime. *Opt. Express* **2022**, *30*, 18300. [CrossRef]
38. Ma, D.Y.; Lu, J.X.; Fang, X.J.; Yang, K.; Wang, K.; Zhang, N.; Han, B.C.; Ding, M. Parameter modeling analysis of a cylindrical ferrite magnetic shield to reduce magnetic noise. *IEEE Trans. Ind. Electron.* **2022**, *69*, 991. [CrossRef]
39. Budker, D.; Gawlik, W.; Kimball, D.F.; Weis, A. Resonant nonlinear magneto-optical effects in atoms. *Rev. Mod. Phys.* **2002**, *74*, 1153. [CrossRef]
40. Lu, Y.; Zhai, Y.; Fan, W.; Zhang, Y.; Xing, L.; Jiang, L.W.; Quan, W. Nuclear magnetic field measurement of the spin-exchange optically pumped noble gas in a self-compensated atomic comagnetometer. *Opt. Express* **2020**, *28*, 17683. [CrossRef]

photonics

MDPI

Communication

Suppression of the Equivalent Magnetic Noise Caused by Electron Spin Polarization in a Xe Isotope Comagnetometer

Zekun Wu [1,2], Zhen Chai [1,2,*], Lan Xiao [1,2] and Zhanchao Liu [1,2,*]

1 School of Instrumentation and Optoelectronic Engineering, Beihang University, Beijing 100191, China
2 Hefei National Laboratory, Hefei 230088, China
* Correspondence: zhenchai@buaa.edu.cn (Z.C.); liuzc@buaa.edu.cn (Z.L.)

Abstract: The Xe isotope comagnetometer in the nuclear magnetic resonance regime can be used as a promising high-precision inertial measurement unit because of the absolute frequency measurement and high bandwidth. The fluctuation of the electron spin polarization leads to equivalent magnetic noise in the Xe isotope comagnetometer, which is one of the main factors limiting the stability of the comagnetometer. Here, we demonstrate systematic research of equivalent magnetic noise suppression and analyze the influence of the electron spin polarization on the Xe isotope comagnetometer. Based on the spin–exchange method between Xe isotopes and alkali metal atoms through the Fermi contact hyperfine interaction, the error equation of the Xe Larmor frequency is established. The equivalent magnetic noise can be suppressed by controlling the static magnetic field. This suppression method for Xe isotope comagnetometers improved the stability while maintaining high bandwidth. The experimental results show that this method can reduce the fluctuations of the ^{129}Xe and ^{131}Xe frequencies by 75% and 68.6%, respectively.

Keywords: Xe isotope comagnetometer; electron spin polarization; noise suppression; equivalent magnetic noise

1. Introduction

Atomic comagnetometers using optically pumped alkali metal atoms to polarize noble gas molecules have a great advantage of having the long coherence time of the nuclear spin [1–3]. Benefiting from the absolute frequency measurement and high bandwidth, comagnetometers in the nuclear magnetic resonance (NMR) regime are widely used in searching for axion-like dark matter [4–6] and the inertial measurement [7,8]. Moreover, the system of Xe isotopes mixed with alkali metal atoms is a universal choice to suppress the effects of the ambient stray field [9–11]. The collisions of the Xe isotopes with polarized alkali metal atoms result in an effective magnetic field proportional to the electron spin polarization, which can shift the Xe Larmor frequency [12,13]. Therefore, the equivalent magnetic noise caused by the fluctuation of the electron spin polarization is undesired for the stability of the Xe isotope comagnetometer.

The electron spin polarization is mainly affected by the temperature, intensity, and frequency of the pump beam [14,15]. Many researchers suppressed the equivalent magnetic noise by only controlling a single factor, such as the stability of the pump beam frequency based on the light absorption method [16,17], the stability of the pump beam intensity relying on additional instruments [15,18], and the optimal temperature [19–21]. They lacked an overall analysis of the equivalent magnetic noise and failed to consider the electron spin polarization as a whole.

In this study, we demonstrate systematic research of equivalent magnetic noise suppression, and study the influence of the electron spin polarization on the Xe isotope comagnetometer operated in the NMR regime. We describe the spin exchange between Xe isotopes and alkali metal atoms through the Fermi contact hyperfine interaction, and then obtain the error equation of the Xe Larmor frequency. The absolute electron spin polarization in different average photon spins is measured based on the free induction decay (FID) method to assess the magnitude of the equivalent magnetic noise. By analyzing the factors affecting the Xe Larmor frequency, we found that the equivalent magnetic noise can be suppressed by controlling the static magnetic field. The long-term stability of the Xe isotope comagnetometer is measured in the open-loop control and closed-loop control. Additionally, the impact of this method on the bandwidth is experimentally studied.

2. Principle

In the vapor cell of a Xe isotope comagnetometer, the nuclear spins are polarized by the spin–exchange interaction with alkali metal atoms that are optically pumped by the circularly polarized light. Neglecting the interaction between the neighboring nuclei, the dynamics of the comagnetometer with nuclear magnetization **K** can be described using the Bloch equation as [22]:

$$\frac{d\mathbf{K}}{dt} = -\omega \times \mathbf{K} - \{\Gamma_2^n, \Gamma_2^n, \Gamma_1^n\} \cdot \mathbf{K} + \mathbf{R_{se}} \tag{1}$$

where $\omega = \gamma_n \mathbf{B}$ is the effective Larmor frequency of the nuclear spins related to the applied magnetic field, the effective magnetic field B_e, and the inertial rotation ω_R. γ_n is the gyromagnetic ratio. The separate matrix term accounts for the longitudinal relaxation rate Γ_1^n and transverse relaxation rate Γ_2^n. Additionally, $\mathbf{R_{se}}$ is determined by the spin–exchange interaction between noble gas molecules and polarized alkali metal atoms. The effective magnetic field of the alkali metal atoms due to the Fermi contact interaction can be expressed as [23,24]:

$$B_e = -\kappa \frac{8\pi g_s \mu_B}{3} n_e P_e \tag{2}$$

where $g_s \approx 2$, μ_B is the electron magnetic moments. n_e is the density of the alkali metal atoms. κ is the enhancement factor for the Rb-Xe ensemble, which varies in amplitude for the Xe isotopes [13]. The electron spin polarization P_e can be expressed as [25]:

$$P_e = s \frac{R_{op}}{R_{op} + R_{rel}} \tag{3}$$

where the alkali metal atoms are polarized by the pump beam at the pump rate R_{op} and the average photon spin s. The alkali relaxation rate R_{rel} exists due to the volume of the cell and the motion of atoms and molecules [26,27]. The optical pumping rate is related to the intensity and frequency of the pumping beam fitting Voigt profiles, which can be expressed as [27]:

$$R_{op} = \frac{I_0}{h v r^2} r_e c f \sum_{F,F'} A_{F,F'} \mathrm{Re}\left[\mathrm{V}\left(\mathrm{vs.} - \mathrm{v}_{\mathrm{F,F'}}\right)\right] \tag{4}$$

where I_0 is the pumping intensity and r is the spot radius, h is the Plank constant and v is the frequency of the pumping beam, r_e is the classical electron radius, c is the light velocity, f is the oscillator strength, and $A_{F,F'}$ is the transition strength. In this case, the drift of the pumping beam intensity and frequency will cause instability for the electron spin polarization, which can cause the equivalent magnetic noise δB_e. According to Equation (3), the electron spin polarization changes linearly with the average photon spin. When alkali metal atoms are pumped by the light of the ellipticity $\frac{\pi}{4} - \alpha$, the average photon spin s is provided by $s = \cos 2\alpha$ [25,28] (define that α is the angle of the fast axis of the quarter-wave plate with respect to the polarization axis of the polarized beam splitter). It is possible to suppress the equivalent magnetic noise by controlling the ellipticity of the pumping beam, but this will adversely affect the spin–exchange rate $\mathbf{R_{se}}$.

The Larmor frequency of Xe isotopes with the equivalent magnetic noise can be expressed as:

$$\omega_a = \gamma_a (B_0 + B_e^a + \delta B_e^a) + \omega_R \tag{5}$$

$$\omega_b = -\gamma_b\left(B_0 + B_e^b + \delta B_e^b\right) - \omega_R \quad (6)$$

where B_0 is the static magnetic field applied along the pump beam. The subscript or superscript a and b correspond to ^{129}Xe and ^{131}Xe, respectively. Besides that, the Xe isotopes respond oppositely to the inertial rotation since their nuclear spins have opposite signs. The inertial rotation can be eliminated by adding the Larmor frequencies of the Xe isotopes, which can be expressed as:

$$\omega_{sum} = (\gamma_a - \gamma_b)B_0 + \gamma_a B_e^a\left(1 - \frac{\gamma_b}{\gamma_a}\Delta B_e\right) \quad (7)$$

where ΔB_e is the equivalent magnetic noise. The high-order noise can be omitted, which is far less than the equivalent magnetic field. Alternatively, the static magnetic field B_0 can be controlled to suppress the equivalent magnetic noise. In this way, when the Larmor frequencies of the Xe isotopes are summarized, the sum frequency only corresponds to the magnetic field and is unaffected by the inertial rotation. It can be seen from Equation (7) that the sum frequency contains two parts: the static magnetic field part and the equivalent magnetic noise part. Both changing B_0 and δB_e can adjust the magnetic field along the z-axis to change the precession frequency of the Xe isotopes. Therefore, by controlling B_0 to compensate δB_e, the sum frequency can keep the stability. We designed a closed-loop control of the static magnetic field to suppress the influence of the equivalent magnetic noise on the precession frequency stability for both ^{129}Xe and ^{131}Xe.

3. Experimental Setup

Figure 1 provides the schematic of a Xe isotope comagnetometer. The vapor cell with a 3 mm inner-side length is heated to 120 °C by two homemade polyimide heaters, filled with 5 Torr ^{129}Xe, 20 Torr ^{131}Xe, 20 Torr N_2, 200 Torr ^{4}He, and a small drop of Rb. A three-dimensional Helmholtz coil designed in our system produces uniform magnetic fields and compensates for the residual magnetism. The static magnetic field along the z-axis is generated using an ultra-low noise current source whose noise density is 4 nA/$\sqrt{\text{Hz}}$ (Stanford Research Systems CS580). The magnetic shield is set up outside the coils to weaken the effect of an external magnetic field. The pump beam with the frequency at the Rb D1 line first passes through the lens to acquire a 2.5 mm diameter. By rotating the half-wave plate, the intensity of the pump beam after the polarized beam splitter is set to 2 mW and linear polarization is ensured. A quarter-wave plate establishes the pump beam at a circular polarization to optically pump the alkali metal atoms. By using a neutral density filter, the probe beam is adjusted to 1 mW. The pump and probe laser are both emitted by the same type of DBR laser diodes (Photodigm PH795DBR080TS) and are driven by the circuit produced by UniQuanta. Passing through the vapor cell, the probe beam is separated into two mutually perpendicular linearly polarized beams using a polarized beam splitter. The signal of the comagnetometer is obtained by a balanced photodetector (Thorlabs PDA015A) and is processed in a lock-in amplifier (Zurich Instruments HF2LI). The concrete operation process can be accomplished in PC software (LabOne), which is applied by Zurich Instruments. As a result of the demodulation and filtering of the original signal from the balanced photodetector, the real-time Larmor frequencies of Xe isotopes can be obtained. The nuclear response curves can be measured by sweeping the frequency of the oscillating magnetic field along the x-axis. The resonant frequency is the Larmor frequency of the Xe isotope. The sum frequency (ω_a^{ref}+ω_b^{ref}) is regarded as the reference frequency. By comparing the real-time frequency and the reference frequency, the frequency discriminator calculates the difference between them and inputs it into the PID controller. It is necessary to ensure that the real-time output frequency of the nucleus is equal to the set value. The driving current consists of the output signal of the PID controller and the DC current input to the z-axis coil to control the static magnetic field.

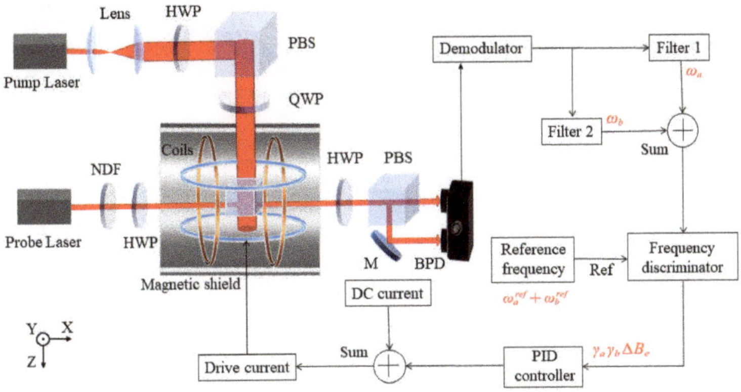

Figure 1. The schematic of a Xe isotope comagnetometer and processing. HWP: half-wave plate, QWP: quarter-wave plate, PBS: polarized beam splitter, PD: photodetector, NDF: neutral density filter, BPD: balanced photodetector,and Ref: reference signal used in the frequency discriminator.

4. Results and Discussion

To assess the magnitude of the equivalent magnetic noise, the electron spin polarization is measured by the nuclear frequency shift $\Delta\nu_n$, which is widely used in the spin-exchange optical pumping system [14,15,29]. The equivalent magnetic field produced by electron spin polarization can be calculated by $\Delta\nu_n = \gamma_n B_e$. Firstly, the alkali metal atoms are pumped by a circularly polarized beam. A $\pi/2$ pulsed magnetic field is applied along the x-axis to drive the ^{129}Xe precession and the coherence decays naturally. Then, the pump beam is blocked before the next $\pi/2$ pulsed magnetic field is applied. An example set for free induction decay (FID) signal and its Fast Fourier Transform (FFT) is shown in Figure 2. The black line is measured with the optical pump condition, and the red line is without the laser irradiation. The nuclear frequency shift is 0.191 Hz corresponding to a 16.1 nT equivalent magnetic field.

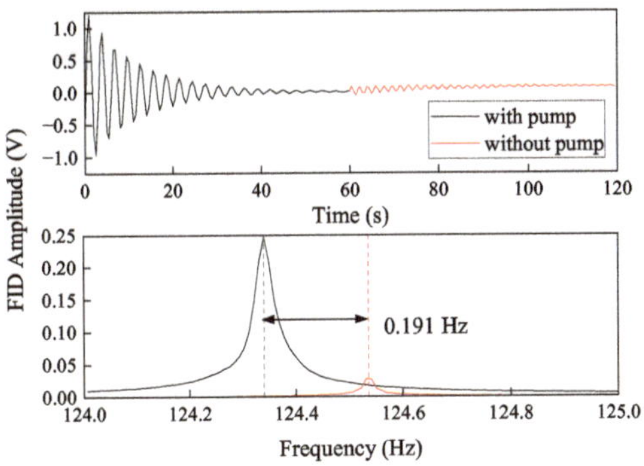

Figure 2. Top: FID signal of ^{129}Xe with and without a circular pump beam. Bottom: FFT of transients.

The relationship between the electron spin polarization and the average photon spin can be obtained by rotating the QWP, while the other operations remain unchanged. We calculated the electronic spin polarization of the simulation results and compared it with the experiment as shown in Figure 3. The measured results are also based on the nuclear frequency shift method as described above. As the average photon spin changes from 1 to 0.18 (corresponding to α from 0° to 40°), the electronic spin polarization decreases gradually from 0.54 to 0.2. This indicates a change in the equivalent magnetic field of 5.96 nT. However, the signal-to-noise ratio of the comagnetometer will be affected by changing the average photon spin. The low electron spin polarization is able to cause a false signal and influence the system sensitivity.

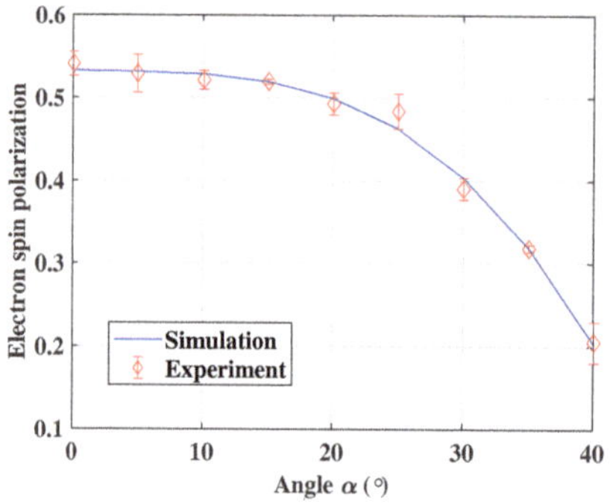

Figure 3. Comparison of the experimental results with simulations of the electron spin polarization in different ellipticity. The error bars represent the maximum and minimum values in the repeated experiments.

In the following, we study the suppression of the equivalent magnetic noise by controlling the static magnetic field. The reference frequency is acquired from the analysis of the frequency response for different magnitudes of the static magnetic field as shown in Figure 4. The magnetic linewidth is the transverse relaxation rate that consists of the relaxation in alkali metal collisions, the relaxation due to collisions with the cell wall surface, the relaxation due to magnetic-field inhomogeneities, and the gas-phase relaxation due to the self-collisions of the noble gas atoms [30,31]. Each of the blue, pink, green, and yellow dots and lines corresponds to the static magnetic field B_0 at 10.79 μT, 10.82 μT, 10.85 μT, and 10.88 μT. Figure 4a shows that the Larmor precession frequency of ^{129}Xe changes from 127.430 Hz to 128.483 Hz. Likewise, the Larmor precession frequency of ^{131}Xe changes proportionally, as shown in Figure 4b. With the increasing magnitude of the static magnetic field, the Larmor frequency of Xe isotopes rises, which is consistent with Equations (5) and (6).

To further verify the linear relationship described in Equations (5) and (6), the Larmor frequencies of the Xe isotopes are measured at a series of static magnetic fields, as shown in Figure 5. The coefficient of determining the R-Square of the Xe isotopes' experimental data are both larger than 98.6%, which describes a good linearity of the static magnetic field and the Larmor frequency of the Xe isotopes. The inset figure represents the ratio of the two frequencies under different magnitudes of the static magnetic field. The ratio is a constant when the equivalent magnetic noises equal to zero. When the equivalent magnetic noise cannot be ignored, the ratio tends to maintain a steady level with the growing of B_0, thus indicating the synchronization and stability of the inertial information in the Xe isotope measurements.

The theoretical value marked in the solid line approximates the ratio of the spin magnetic ratios of the Xe isotopes. In addition, it is easier to control the static magnetic field than the average photon spin in the experimental operations. Therefore, controlling the static magnetic field is a better way to suppress the equivalent magnetic noise caused by electron spin polarization.

The real-time Larmor frequencies of Xe isotopes in the static magnetic field of open-loop control and closed-loop control are measured over 15 h, as shown in Figure 6. The open-loop control is conducted in the first 9 h, in which the range of frequency fluctuations of ^{129}Xe and ^{131}Xe, due to the equivalent magnetic noise, is 0.01 Hz and 0.0035 Hz, respectively. The closed-loop control conducted in the last 6 h is applied to suppress the long-term drift caused by this noise. The variation in the frequency fluctuations of ^{129}Xe and ^{131}Xe are 0.0025 Hz and 0.0011 Hz, respectively, reduced by 75% and 68.6%. This notably reduces the interference of the noise and improves the stability of the Xe isotope comagnetometer. A greater enhancement of the noise suppression effect is restricted by magnetic field gradient produced by inhomogeneous electron spin polarization and errors introduced by signal processing. Figure 7 shows the signal noise measurement results of the Xe isotope comagnetometer under the same conditions of Figure 6. According to the experimental test results, the signal noise of the co-magnetometer in closed-loop control is lower than it in the open-loop control. Both frequencies of ^{129}Xe and ^{131}Xe have better performance under the closed-loop control. The noise level of ^{129}Xe is greater than ^{131}Xe because of the larger gyromagnetic ratio.

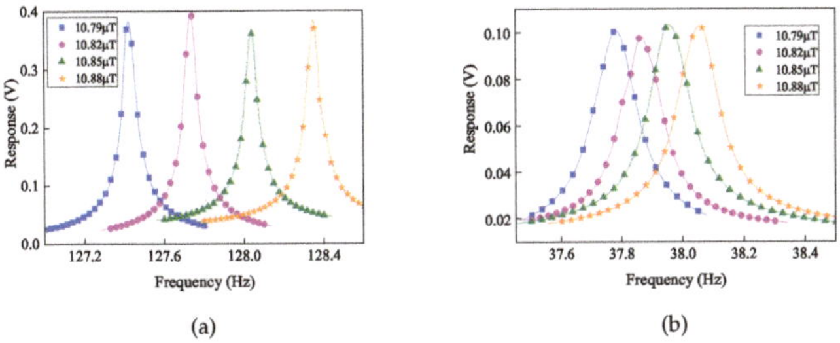

Figure 4. Xe isotopes response amplitudes (**a**) ^{129}Xe, (**b**) ^{131}Xe under different magnitudes of the static magnetic field. B_0 changes from 10.79 μT to 10.88 μT.

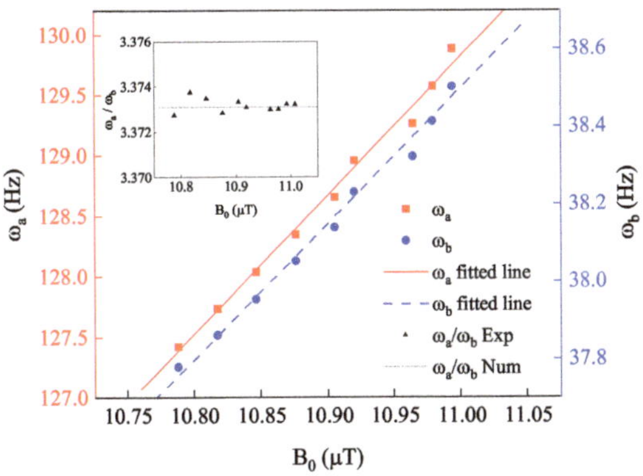

Figure 5. Measured nuclei Larmor frequency at different static magnetic fields. In the legend, the subscript "a" denotes ^{129}Xe, and subscript "b" denotes ^{131}Xe. Inset: dependence of the ratio of precession frequencies of ^{129}Xe and ^{131}Xe.

To assess the influence of this method on the bandwidth, the output frequency of the Xe isotope comagnetometer is measured as shown in Figure 8. As can be seen from Equations (5) and (6), the output frequency is provided by:

$$\omega_R = \frac{\gamma_a \omega_b + \gamma_b \omega_a}{\gamma_b - \gamma_a} \tag{8}$$

The input rotation rate changes from 300 °/s to −300 °/s at a step of 50 °/s, and Figure 8a shows the absolute frequency measured using the comagnetometer with the sampling rate at 200 Hz. The output frequencies calculated according to the standard that the method in each rotation rate (circle) are compared with the input signal (line), which can be seen in Figure 8b. The scale factor is 0.997. Thereby, our method to suppress the equivalent magnetic noise caused by electron spin polarization has realized a groundbreaking compromise between the stability and bandwidth.

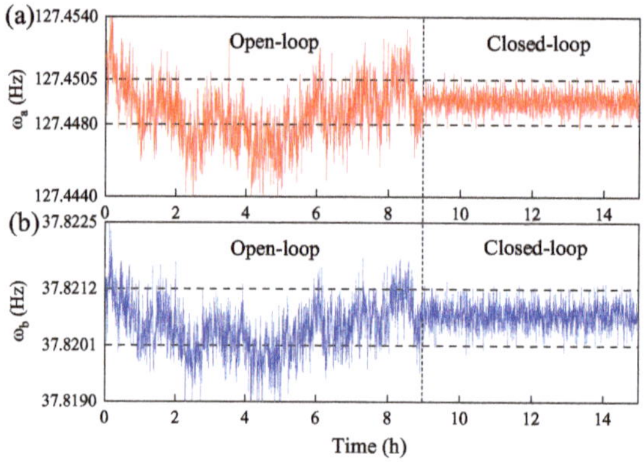

Figure 6. Experimental performance of the Xe isotopes under open-loop control and closed-loop control. (**a**) relates to ^{129}Xe, while (**b**) relates to ^{131}Xe. The frequencies of the Xe isotopes are measured simultaneously. The horizontal dashed lines indicate the noise level of the closed-loop control.

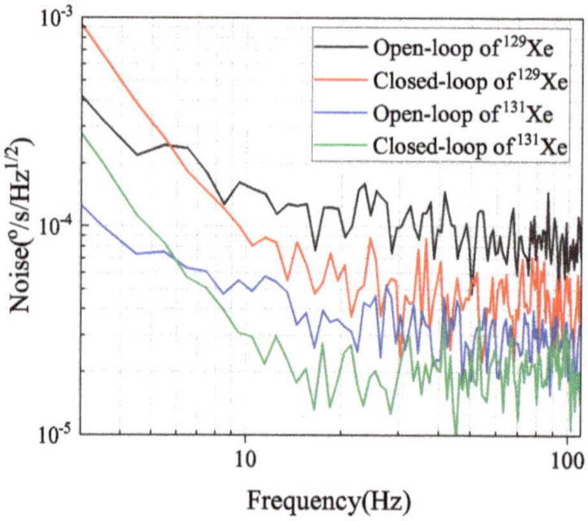

Figure 7. Comagnetometer signal noise of ^{129}Xe and ^{131}Xe under open-loop control and closed-loop control.

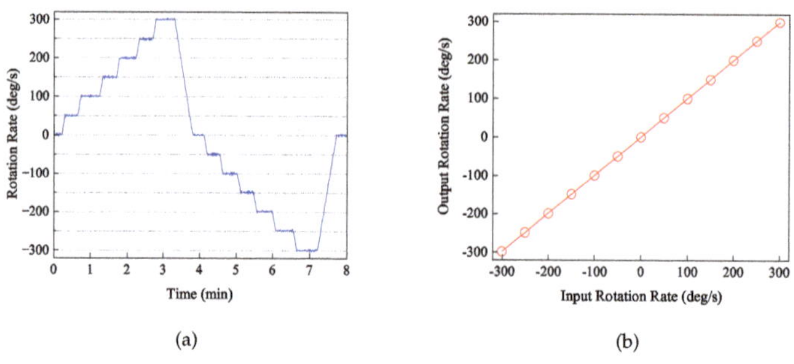

Figure 8. The output frequency ω_R of the Xe isotope comagnetometer in rotation range from 300 °/s to −300 °/s. (**a**) The output rotation rate in the time domain. (**b**) The relationship between input (line) and output (circle) rotation rate.

5. Conclusions

In conclusion, the influence of the electron spin polarization on the Larmor frequency of the Xe isotope comagnetometer was analyzed in this study. The drift of the pumping beam intensity and frequency causes fluctuation for the electron spin polarization, which can lead to the equivalent magnetic noise. We found that the equivalent magnetic noise can be suppressed by controlling the static magnetic field rather than adjusting the ellipticity of the pump beam, which can improve the stability while maintaining a high bandwidth. Experimentally, the long-term stability of the Xe isotope comagnetometer is measured under the open-loop and closed-loop operations. The fluctuations of the ^{129}Xe and ^{131}Xe frequencies have been reduced by 75% and 68.6%, respectively. This study provides a novel method to suppress the equivalent magnetic noise caused by electron spin polarization and has great significance in the performance of the Xe isotope comagnetometer.

Author Contributions: Conceptualization, Z.W. and Z.C.; methodology, Z.W.; software, Z.L.; validation, Z.W. and L.X.; formal analysis, Z.W.; investigation, Z.C.; resources, Z.C.; data curation, Z.W.; writing—original draft preparation, Z.W.; writing—review and editing, Z.C. and Z.L.; visualization, Z.W.; supervision, Z.L.; project administration, Z.L.; funding acquisition, Z.L. All authors have read and agreed to the published version of the manuscript.

Funding: This research was funded by Innovation Program for Quantum Science and Technology (2021ZD0300403) and National Science Fund for Distinguished Young Scholars (62225102).

Institutional Review Board Statement: Not applicable.

Informed Consent Statement: Not applicable.

Data Availability Statement: The data presented in this study are available from the corresponding author upon reasonable request.

Conflicts of Interest: The authors declare no conflict of interest.

References

1. Meyer, D.; Larsen, M. Nuclear magnetic resonance gyro for inertial navigation. *Gyroscopy Navig.* **2014**, *5*, 75–82. [CrossRef]
2. Korver, A.; Thrasher, D.; Bulatowicz, M.; Walker, T. Synchronous spin-exchange optical pumping. *Phys. Rev. Lett.* **2015**, *115*, 253001. [CrossRef] [PubMed]
3. Walker, T.; Larsen, M. Spin-exchange-pumped nmr gyros. In *Advances in Atomic, Molecular, and Optical Physics*; Elsevier: Amsterdam, The Netherlands, 2016; pp. 373–401.
4. Afach, S.; Buchler, B.; Budker, D.; Dailey, C.; Derevianko, A.; Dumont, V.; Figueroa, N.; Gerhardt, I.; Grujic, Z.; Guo, H.; et al. Search for topological defect dark matter with a global network of optical magnetometers. *Nat. Phys.* **2021**, *17*, 1396–1401. [CrossRef] [PubMed]
5. Jiang, M.; Su, H.; Garcon, A.; Peng, X.; Budker, D. Search for axion-like dark matter with spin-based amplifiers. *Nat. Phys.* **2021**, *17*, 1402–1407. [CrossRef]
6. Jackson, D.; Afach, S.; Aybas, D.; Blanchard, J.; Budker, D.; Centers, G.; Engler, M.; Figueroa, N.; Garcon, A.; Graham, P.; et al. Overview of the cosmic axion spin precession experiment. In *Microwave Cavities and Detectors for Axion Research*; Springer: Berlin/Heidelberg, Germany, 2020; pp. 105–121.
7. Bevan, D.; Bulatowicz, M.; Clark, P.; Flicker, J.; Griffith, R.; Larsen, M.; Luengo-Kovac, M.; Pavell, J.; Rothballer, A.; Sakaida, D. Nuclear magnetic resonance gyroscope: Developing a primary otation sensor. In Proceedings of the 2018 IEEE International Symposium on Inertial Sensors and Systems (INERTIAL), Lake Como, Italy, 26–29 March 2018; pp. 1–2.
8. Cipolletti, R.; Riedrich-Moeller, J.; Fuchs, T.; Wickenbrock, A.; Budker, D. Modeling of the transient behavior of a nuclear magnetic resonance gyroscope. In Proceedings of the 2021 IEEE Sensors, Sydney, Australia, 31 October–3 November 2021; pp. 1–4.
9. Hao, C.; Yuan, C.; Liu, S.; Sheng, D. Herriott-cavity-assisted closed-loop Xe isotope comagnetometer. *Phys. Rev. A* **2021**, *103*, 053523. [CrossRef]
10. Lee, S.; Yim, S.; Kim, T.; Kim, Z.; Shim, K. Lock-in-detection in 87Rb–129Xe/131Xe atom spin gyroscopes. *J. Phys. B-At. Mol. Opt. Phys.* **2020**, *53*, 035502. [CrossRef]
11. Zhang, K.; Zhao, N.; Wang, Y. Closed-loop nuclear magnetic resonance gyroscope based on Rb-Xe. *Sci. Rep.* **2020**, *10*, 1–7. [CrossRef]
12. Petrov, V.; Pazgalev, A.; Vershovskii, A. Isotope shift of nuclear magnetic resonances in 129Xe and 131Xe caused by spin-exchange pumping by alkali metal atoms. *IEEE Sens. J.* **2019**, *20*, 760–766. [CrossRef]

13. Bulatowicz, M.; Griffith, R.; Larsen, M.; Mirijanian, J.; Fu, C.; Smith, E.; Snow, W.; Yan, H.; Walker, T. Laboratory search for a long-range t-odd, p-odd interaction from axionlike particles using dual-species nuclear magnetic resonance with polarized 129Xe and 131Xe gas. *Phys. Rev. Lett.* **2013**, *111*, 102001. [CrossRef]
14. Liu, X.; Luo, H.; Qu, T.; Yang, K.; Ding, Z. Measuring the spin polarization of alkali-metal atoms using nuclear magnetic resonance frequency shifts of noble gases. *AIP Adv.* **2015**, *5*, 107119. [CrossRef]
15. Jia, Y.; Liu, Z.; Zhou, B.; Liang, X.; Wu, W.; Peng, J.; Ding, M.; Zhai, Y.; Fang, J. Pump beam influence on spin polarization homogeneity in the nuclear magnetic resonance gyroscope. *J. Phys. D-Appl. Phys.* **2019**, *52*, 355001.
16. Jia, Y.; Liu, Z.; Chai, Z.; Liang, X.; Wu, W. The optimization and stabilization of pump light frequency in the minimized atomic magnetometer. *IEEE Trans. Instrum. Meas.* **2021**, *70*, 1–9. [CrossRef]
17. Dudzik, G.; Rzepka, J.; Abramski, K. Polarization switching detection method using a ferroelectric liquid crystal for dichroic atomic vapor laser lock frequency stabilization techniques. *Appl. Opt.* **2015**, *54*, 2806–2813. [CrossRef] [PubMed]
18. Huang, J.; Fan, W.; Wang, Z.; Yuan, L.; Zhang, K.; Pei, H.; Pang, H.; Quan, H. Analysis and suppression of the misalignment error for the pumping laser in the atomic comagnetometer. *Opt. Express* **2022**, *30*, 6374–6387. [CrossRef]
19. Yin, Y.; Zhou, B.; Yin, K.; Wang, Y.; Tang, J.; Ye, M.; Ning, X.; Han, B. The influence of temperature and modulated magnetic field on the transmission intensity of atomic magnetometer. *J. Phys. D-Appl. Phys.* **2021**, *54*, 485001. [CrossRef]
20. Zhang, S.; Lu, J.; Ye, M.; Zhou, Y.; Yin, K.; Lu, F.; Yan, Y.; Zhai, Y. Optimal operating temperature of miniaturized optically pumped magnetometers. *IEEE Trans. Instrum. Meas.* **2022**, *71*, 1–7. [CrossRef]
21. Lee, W.; Lucivero, V.; Romalis, M.; Limes, M.; Foley, E.; Kornack, T. Heading errors in all-optical alkali-metal-vapor magnetometers in geomagnetic fields. *Phys. Rev. A* **2021**, *103*, 063103. [CrossRef]
22. Grebenkov, D. Nmr survey of reflected brownian motion. *Rev. Mod. Phys.* **2007**, *79*, 1077. [CrossRef]
23. Walker, T.; Happer, W. Spin-exchange optical pumping of noble-gas nuclei. *Rev. Mod. Phys.* **1997**, *69*, 629. [CrossRef]
24. Ma, Z.; Sorte, E.; Saam, B. Collisional 3He and 129Xe frequency shifts in Rb–noble-gas mixtures. *Phys. Rev. Lett.* **2011**, *106*, 193005. [CrossRef]
25. Seltzer, S. *Developments in Alkali-Metal Atomic Magnetometry*; Princeton University: Princeton, NJ, USA, 2008.
26. Happer, W.; Miron, E.; Schaefer, S.; Schreiber, D.; Van Wijngaarden, W.; Zeng, X. Polarization of the nuclear spins of noble-gas atoms by spin exchange with optically pumped alkali-metal atoms. *Phys. Rev. A* **1984**, *29*, 3092. [CrossRef]
27. Kornack, T. *A Test of CPT and Lorentz Symmetry Using a Potassium-Helium-3 Co-Magnetometer*; Princeton University: Princeton, NJ, USA, 2005.
28. Ding, Z.; Yuan, J.; Wang, Z.; Lu, G.; Luo, H. Optically pumped rubidium atomic magnetometer with elliptically polarized light. *Optik* **2016**, *127*, 5270–5273. [CrossRef]
29. Schaefer, S.; Cates, G.; Chien, T.; Gonatas, D.; Happer, W.; Walker, T. Frequency shifts of the magnetic-resonance spectrum of mixtures of nuclear spin-polarized noble gases and vapors of spin-polarized alkali-metal atoms. *Phys. Rev. A* **1989**, *39*, 5613. [CrossRef] [PubMed]
30. Yan, Y.; Lu, J.; Zhang, S.; Lu, F.; Yin, K.; Wang, K.; Zhou, B.; Liu, G. Three-axis closed-loop optically pumped magnetometer operated in the SERF regime. *Opt. Express* **2022**, *30*, 18300–18309. [CrossRef] [PubMed]
31. Liu, X.; Chen, C.; Qu, T.; Yang, K.; Luo, H. Transverse spin relaxation and diffusion-constant measurements of spin-polarized 129Xe nuclei in the presence of a magnetic field gradient. *Sci. Rep.* **2016**, *6*, 1–8. [CrossRef] [PubMed]

photonics

MDPI

Article

Laser Heating Method for an Alkali Metal Atomic Cell with Heat Transfer Enhancement

Yang Li, Guoqing Zhou, Shencheng Tian, Xuejing Liu *, Xiangmei Dong and Xiumin Gao

School of Optical-Electrical and Computer Engineering, University of Shanghai for Science and Technology, Shanghai 200093, China
* Correspondence: liuxuejing@usst.edu.cn

Abstract: Alkali metal atomic cells are crucial components of atomic instruments, such as atomic magnetometers, atomic gyroscopes, and atomic clocks. A highly uniform and stable heating structure can ensure the stability of the alkali metal atom density. The vapor cell of an atomic magnetometer that uses laser heating has no magnetic field interference and ease of miniaturization, making it superior to hot air heating and AC electric heating. However, the current laser heating structure suffers from low heating efficiency and uneven temperature distribution inside the vapor cell. In this paper, we designed a non-magnetic heating structure based on the laser heating principle. We studied the temperature distribution of the heating structure using the finite element method (FEM) and analyzed the conversion and transfer of laser energy. We found that the heat conduction between the vapor cell and the heating chips (colored filters) is poor, resulting in uneven temperature distribution and low heating efficiency in the vapor cell. Therefore, the addition of graphite film to the four surfaces of the vapor cell was an important improvement. This addition helped to balance the temperature distribution and improve the conduction efficiency of the heating structure. It was measured that the power of the heating laser remained unchanged. After the addition of the graphite film, the temperature difference coefficient (CVT) used to evaluate the internal temperature uniformity of the vapor cell was reduced from 0.1308 to 0.0426. This research paper is crucial for improving the heating efficiency of the non-magnetic heating structure and the temperature uniformity of the vapor cell.

Keywords: SERF magnetometer; non-magnetic laser heating; graphite film

1. Introduction

With the rapid development of atomic operation technology and optical technology, the ultra-sensitive magnetic field sensor based on SERF (Spin-Exchange Relaxation-Free) effect, which uses atoms as sensitive components, has been widely used in various fields, such as medicine, national defense, and aerospace. Examples of their applications include magnetoencephalography (MEG) for brain imaging [1–4], magnetocardiography (MCG) for heart imaging [5–7], and geological dating investigations [8]. The world's most sensitive SERF atomic magnetometer has a sensitivity of 160 aT/Hz1/2, which is also the record for magnetic field measurements [9]. In high-sensitivity magnetometers, a drop of alkali metal in the vapor cell needs to be heated into high-density vapor, and the saturation vapor pressure of alkali metal atoms directly affects the sensitivity of the SERF atomic magnetometer. The spatial distribution of alkali vapor density in a large vapor cell has a direct impact on the processing of measurement signals for multichannel magnetic fields. The density distribution depends on the spatial temperature distribution [10]. Therefore, it is essential to optimize the heating method and structure of the vapor cell to improve its temperature stability and uniformity. Additionally, any additional magnetic noise should be avoided during the heating process [11].

There are three common heating methods: electric heating [12–16], flow hot air heating [17,18], and laser heating [19–25]. Although flow hot air heating can ensure that no additional magnetic noise is introduced, its heating efficiency is low, and the overall structure is too large to be integrated. Electric heating can use a lock-in amplifier to filter out additional magnetic noise in the signal, but the working frequency must remove the heating current frequency, which limits the measurement range of the atomic magnetometer. More importantly, this method cannot completely filter out residual magnetic noise. Although intermittent electric heating can completely avoid additional magnetic noise, the atomic magnetometer does not work during the heating process, and it cannot measure the continuous changes in the magnetic field, directly leading to a decrease in the sensitivity of the atomic magnetometer. In contrast, laser heating is a good option compared to other heating technologies because it does not generate additional magnetic noise. Furthermore, as atomic magnetometers continue to miniaturize and integrate, the energy required to heat chip-level atomic cells to the target temperature is very low. The continuous optimization and updating of the laser heating structure also promote laser heating as the mainstream heating method for alkali metal cells in the future. Therefore, the non-magnetic heating method chosen in this study is laser heating. Preusser et al. [23] used a 915 nm semiconductor laser to directly heat the vapor cell, which was heated to 97 °C with a required laser power of 200 mW. The laser energy cannot be completely absorbed by the vapor cell, and the heating efficiency is limited. Mhaskar et al. [24] and Sheng [25] installed colored filters on both sides of the vapor cell to improve the laser absorption efficiency, but there is still residual energy that has not been utilized. In Sheng's study, two 1550 nm laser beams were used simultaneously to heat the vapor cell, which made the heating structure more complex. Therefore, improving the laser heating efficiency and the spatial uniformity of the temperature inside the vapor cell while simplifying the heating structure is also the key to improving laser heating technology.

In this paper, we study a heating structure for a cell that utilizes laser heating technology, which is to improve the heating efficiency and temperature uniformity of the vapor cell. First, a finite element (FEM) model is established to study the internal temperature of the vapor cell and analyze the laser energy conversion and transfer. To solve the problems of low heating efficiency and uneven temperature distribution inside the vapor cell caused by the poor thermal conductivity of the vapor cell and the heating chips (colored filters), we added a graphite film to the heating structure. To validate the proposed method, the operating power of the heating laser is kept constant, and graphite films of different sizes are added to the heating structure. This is performed to measure the temperature variation inside the vapor cell. The CVT and the temperature error distribution plots of the vapor cell are used to characterize the uniformity of the temperature distribution inside the vapor cell.

2. Analysis and Simulation Methods

2.1. Heat Transfer Analysis

The heating process involves three types of heat transfer: thermal radiation, thermal convection, and thermal conduction. The heating chips convert laser energy to thermal energy and transfer it evenly to the outer wall of the vapor cell through heat conduction. The outer wall of the vapor cell then radiates heat to the gas inside the vapor cell. Simultaneously, the colored filters and the vapor cell emit thermal energy to the surrounding environment through thermal radiation. Eventually, the entire system reaches a dynamic equilibrium state, where heat conduction can be expressed by Fourier's law of heat conduction as follows [26]:

$$\frac{\partial^2 T}{\partial x^2} + \frac{\partial^2 T}{\partial y^2} + \frac{\partial^2 T}{\partial z^2} = \frac{\rho \cdot C_p}{k} \cdot \frac{\partial T}{\partial t}. \tag{1}$$

In Equation (1), T is temperature, t is time, ρ is the density of the object, C_p is constant pressure heat capacity, k is the thermal conductivity. The left side of the equation shows the rate of heat transfer, so heat transfer is only affected by the material properties of the object. The heat radiation can be expressed by the Stefan–Boltzmann equation as follows:

$$q_1 = \varepsilon\sigma A\left(T_{amd}^4 - T^4\right), \quad (2)$$

where q_1 is the heat flow rate, ε is the rate of radiation, σ is Stefan–Boltzmann constant, A is the area of radiation surface, T_{amd} is temperature of the vapor cell's wall, and T is the temperature of the alkali metal gas in the vapor cell. The heat convection can be expressed by Newton cooling equation as follows:

$$q_2 = hA_1(T_{Solid} - T_{Fluid}), \quad (3)$$

where h is the heat transfer coefficient, A_1 is the area of the radiating surface, T_{Solid} is the temperature of solid surface, and T_{Fluid} is the fluid temperature around the solid. When the system finally reaches dynamic stability, the thermal balance equation is obtained as follows:

$$Q = \rho C_p \frac{\partial T}{\partial t} + \rho C_p v \cdot \nabla T + \nabla \cdot (-k \cdot \nabla T), \quad (4)$$

where Q is the total heat input; v is the velocity field in the fluid; ∇ is the Nabla operator, and it is defined as a differential operator; it also has vector form, so ∇T is the changes of temperature in all directions in space; and ρ is the density of the object. The heating chips and the vapor cell are closely bonded and placed in the insulating layer. When the system reaches a stable state, the primary heat transfer mode is conduction, and convection and radiation can be disregarded. In this study, both the vapor cell and the heating chips are made of quartz glass, which has low thermal conductivity and high thermal capacity at constant pressure. Therefore, the thermal conductivity of the system is poor. To improve the thermal conductivity of the system, the graphene thin layers are added to the heating structure. As a material with high thermal conductivity, graphite sheets have a lamellar structure with anisotropy [27–29]. The thermal conductivity is high along the lamellar direction and low in the vertical direction.

2.2. Finite Element Analysis

Based on the COMSOL Multiphysics simulation software, the temperature distribution of a cell heated by a pair of heating chips on the X-axis is studied using the finite element analysis method. The required physical field is heat transfer in solid. The simulation process consists of four steps. First, the heating structure model is built, including heating chips, the vapor cell, and graphite films. Second, material properties are set, and physical fields are added for solid and fluid heat transfer. The heating chip heats the vapor cell by absorbing the energy of the laser, so the boundary of the heating chip is set as the heat source in the way of heat flux. The energy of the heating laser is a continuous laser with a Gaussian spatial distribution (GSD). The laser source is defined as follows [30]:

$$Q = GSD \times \rho_{laser}. \quad (5)$$

Q is the heat produced by the laser; ρ_{laser} is the energy density of the laser, expressed as $2 \cdot P_{laser}/D$; P_{laser} is the working power of the laser; and D is the area of the laser spot. Gaussian spatial distribution (GSD) is represented by Equation (6):

$$GSD = e^{-2 \cdot r_{focus}^2 / r_{spot}^2}. \quad (6)$$

r_{spot} is the radius of the laser spot. r_{focus} is the distance to the center of the laser spot, which is expressed as $r_{focus} = \sqrt{\left(x - x_{focus}\right)^2 + \left(y - y_{focus}\right)^2 + \left(z - z_{focus}\right)^2}$; $\left(x_{focus}, y_{focus}, z_{focus}\right)$ represents the coordinates of the center point of the laser source; the radius of the laser spot is set to 4 mm; and the initial temperature of the environment is set to 20 °C. In order to improve the calculation precision, the whole heating model is partitioned by grid. The sizes of the grid units of the heating chips and the vapor cell are, respectively, set to fine and extremely fine. Finally, the transient temperature of the heating structure is analyzed, and the temperature distribution when temperature of the vapor cell reaches dynamic stability is analyzed.

3. Experimental Setup

The rubidium SERF magnetometer comprise a cell, a pump laser, a detection laser, a magnetically shielded barrel, and a non-magnetic heating structure. The non-magnetic heating structure is responsible for heating the rubidium vapor to operational temperature without creating any magnetic noise during the heating process. In this paper, we selected a laser to heat the rubidium gas cell. As illustrated in Figure 1, the heating structure consists of the following components: a 1550 nm heating laser, two heating chips (colored filters), a temperature sensor (PT100), a cell, an insulation layer, a graphite film, and a thermal imaging camera with a real-time 160 × 120-pixel thermal map.

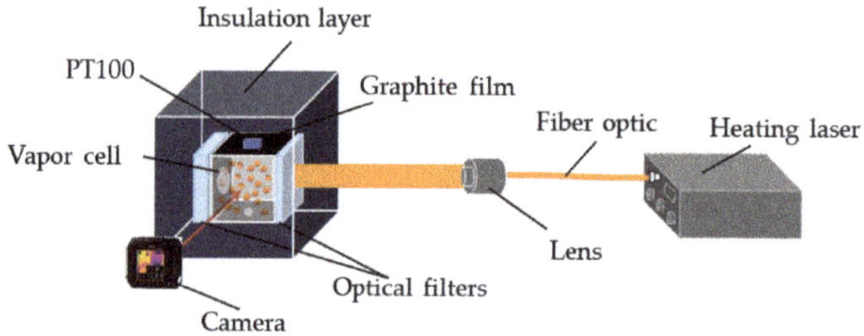

Figure 1. Non-magnetic laser heating structure.

In this experiment, colored filters from SCHOTT were utilized as heating chips to absorb laser energy in the 1550 nm wavelength range. The incident and exit window filters for the heating laser were 1 mm UG-5 and 3 mm RG-9, respectively; the incident optical power absorbed by the two filters was 56.3% and 100%, allowing each filter to absorb approximately equal amounts of heating laser energy; and the converted heat was then used to heat the vapor cell. The temperature and temperature distribution of the vapor cell were measured via a temperature sensor (PT100) and thermal imager, respectively, the rubidium atomic vapor and buffer gas are enclosed in a 20 mm × 20 mm × 20 mm cubic alkali metal cell. An insulating layer is incorporated to reduce heat loss and enhance heat transfer efficiency; the cube-shaped insulating layer with a side length of 40 mm is made of polyimide foam, which has a thermal conductivity of 0.035 W/(m·K). Two cylindrical holes with a diameter of 10 mm intersect with the cube, allowing for the passage of lasers that irradiate the vapor cell.

After conducting the experiment, it was found that the thermal conductivity of the vapor cell and heating was poor. Therefore, a graphite film is added to the heating structure as a high thermal conductive layer to balance the temperature distribution and improve the conduction efficiency. The thickness of the graphite film is 0.2 mm. The thermal conductivity along the fiber axis is 800 W/(m·K), and the thermal conductivity perpendicular to the fiber axis is 10 W/(m·K). The graphene thin layer, as shown in Figure 2, is a square sheet of 20 mm × 80 mm in size. This sheet contains two circular holes of 10 mm diameter each for light passage and one circular hole of 6 mm diameter for passage through the vapor cell shank.

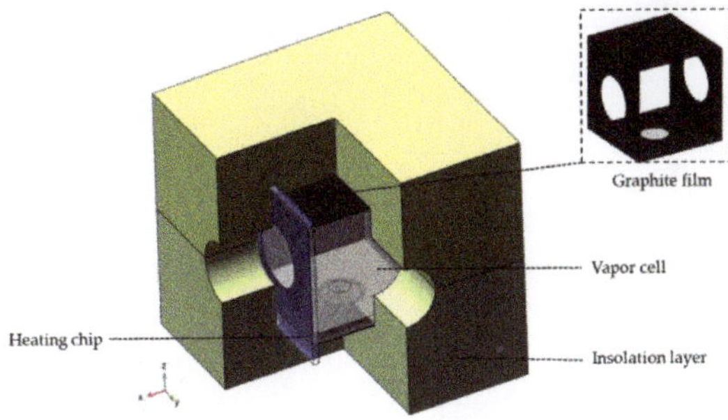

Figure 2. The core part of the heating structure.

4. Results and Discussion

4.1. Simulated Results

During the simulation process, it was observed that the thermal conductivity of the heating chips and the vapor cell was extremely low. As a result, a significant amount of heat was concentrated in the center of the heating chips, and it is not transferred to the vapor cell, which led to a high level of thermal loss and low internal temperature uniformity in the vapor cell. To address this issue, the thin layer structure was introduced for solid heat transfer. A graphite film was added to the contact surface of the vapor cell and the heating chips to increase the thermal conductivity of the heating structure. The layer thickness was set to 0.2 mm for optimal results, and the thermal conductivity along the fiber direction of the graphite film was 800 W/(m·K), whereas that perpendicular to the fiber axis was 10 W/(m·K). Figure 3 displays the distribution of isotherms on the heating chips' surface with or without the addition of a graphite film. The temperature distribution of the heating chips with the graphite film is more uniform compared to the one without the graphite film under the same power of the heating laser.

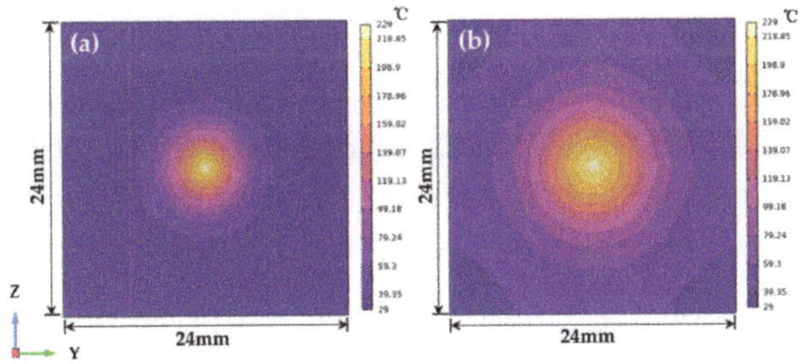

Figure 3. Graphite film reduces the temperature gradient on the surface of the heating chip. (**a**) The distribution of isotherms of the heating chip without adding a graphite film; (**b**) The distribution of isotherms of the heating chip with adding a graphite film.

The ambient temperature is 20 °C, the laser power is set to 2.4 W, and the diameter of the heated laser spot is 8 mm. Heating time can be reduced via a pair of heating chips to heat the vapor cell simultaneously. The temperature gradient of the vapor cell is smaller when a heater is placed in each of the two windows of the heating laser compared to placing a heater in only one window. Using the finite element method to simulate the heat transfer of the heating structure with added graphite film, Figure 4a shows the temperature distribution of the two heating chips when heating the vapor cell.

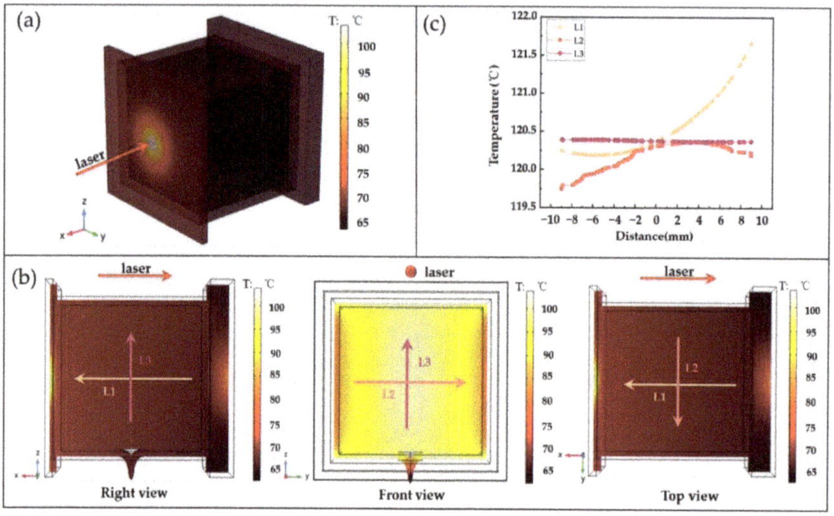

Figure 4. (**a**) The temperature distribution diagram of the heating structure when the temperature in the vapor cell is around 120 °C; (**b**) Temperature distribution on the XAY, YAZ, and XAZ planes; (**c**) Taking the center of the vapor cell as the coordinate origin, the temperature curve along the three axes.

We define the center point of the vapor cell as A, and the space coordinates of point A is (0,0,10), and two heating chips are distributed at 10 mm and −10 mm of the x axis. Figure 4b shows the temperature distribution of the heating structure on the XAY, YAZ, and XAZ planes. L1, L2, and L3 represent three straight lines passing through the x, y, and z axes of the central point inside the vapor cell. In contrast to the temperature graph in Figure 4c, the temperature gradient in the direction of the X-axis is larger. This is because the perforations of the insulating layer are oriented in the direction of both the x-axis and the y-axis, and the two heating chips are also located in the direction of the x-axis. Furthermore, the heating chip located in the incident window of the heating laser absorbed 12.6% more laser energy than the heating chip located in the outgoing window of the heating laser. This is the main reason for the large temperature gradient along the X-axis. The smaller temperature gradient in the Y-axis direction is due to the smaller aperture and less thermal exchange between the vapor cell and the external environment. In the Z-axis direction, the vapor cell's temperature gradient is the smallest, and the outer wall temperature of the vapor cell is close to the interior temperature. Therefore, to reduce the temperature measurement error, the temperature sensor (PT100) can be placed in the upper shell center of the vapor cell to measure its temperature.

4.2. Experiments

To study the uniformity of the vapor cell under laser heating, we measured the internal temperature using a temperature sensor (PT100) and a thermal imager. The temperature sensor (PT100) was attached to the upper center of the vapor cell, and the measurement results are shown in Figure 5.

When graphite film is added to the four sides of the vapor cell, the operating power of the 1550 nm heating laser is 2.4 W. It can be seen from Figure 5 that when the power of the heating laser is 2.4 W, the simulated interior temperature of the vapor cell is 120 °C, while the experimental temperature of the vapor cell is 118 °C. This is because the thermal insulation structure cannot completely eliminate the vapor cell and the outside thermal radiation, resulting in heat loss of the vapor cell, so there is a deviation between theoretical simulation and experimental data.

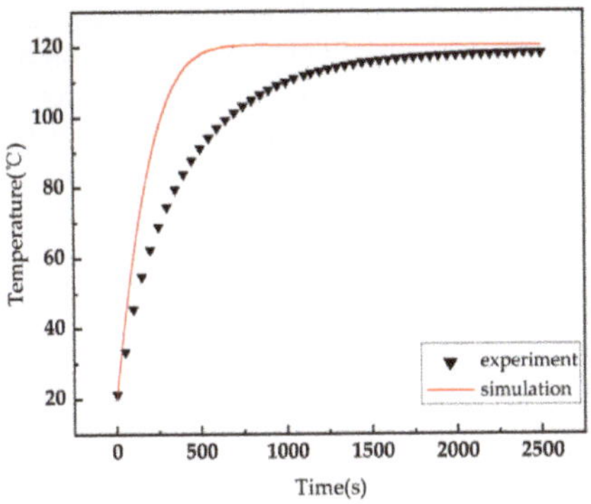

Figure 5. Simulated (red) and experimental (black) temperature curves of the vapor cell were measured.

Figure 6 is a thermal image of the internal temperature radiation of the vapor cell with the *Y*-axis facing. This is when the graphite film is not added to the surface of the vapor cell, and the graphite film with different sizes is added.

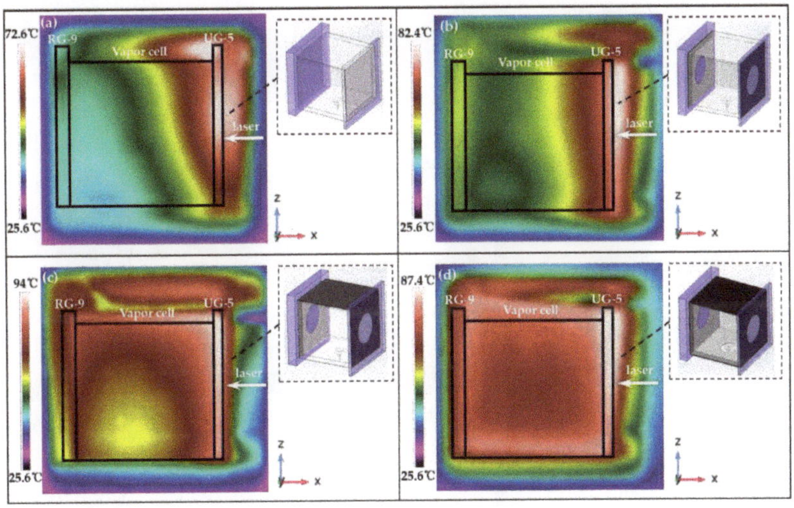

Figure 6. Temperature radiography was performed on the vapor cell with different sizes of graphite films added. (**a**) the vapor cell without any graphite films; (**b**) the vapor cell with graphite films on two surfaces; (**c**) the vapor cell with graphite films on three surfaces; (**d**) the vapor cell with graphite films on four surfaces.

From Figure 6, it is evident that the temperature near the laser incident window in the vapor cell is the highest. This is because the heating chip located in the exit window absorbed too much laser energy, resulting in the two colored filters absorbing different amounts of laser energy. Therefore, one can select either a thin heating chip or a heating chip with a low absorption rate to the heating laser. Figure 6a shows the temperature radiography of the vapor cell without the addition of a graphite film. The material of the vapor cell and the heating chips is quartz glass, which has a poor thermal conductivity at 1.4 W/(m·K). This leads to a significant accumulation of heat in the incident window of the laser. The thermal energy inside the vapor cell comes from the radiant heat of the vapor cell wall. However, there is a shank at the bottom of the vapor cell, which increases the radiant distance at the bottom of the vapor cell. As a result, the temperature of the bottom side is lower than the upper side. In the experiment, when the laser power reached

600 mW, the UG-5 heating chip broke due to the large difference in temperature between the center and the surrounding areas. In Figure 6b, the graphite film was pasted on the two surfaces of the vapor cell that were in contact with the heating chips, which obviously improved the uniformity of the temperature distribution of the incident window. However, the temperature of the exit window remained low, which was due to the poor thermal conductivity of the vapor cell. In Figure 6c, we have pasted graphite films on the upper surface of the vapor cell to increase heat transfer along the *X*-axis. As can be seen from the figure, the heat from the incident window is transferred to the exit window by the graphite film, improving the temperature uniformity of the upper part of the vapor cell.

In order to improve the temperature uniformity of the vapor cell, the graphite film covers the upper and lower top surfaces of the vapor cell, the incident light window, and the outgoing light window at the same time. Under the same heating condition, the thermal radiation Figure 6d is obtained. By comparing the temperature radiation in Figure 6, it can be concluded that the graphite film improves the thermal conductivity of the system and makes the temperature distribution of the vapor cell more uniform. However, the thermal radiation between the vapor cell and the outside also increased, and the overall temperature of the vapor cell decreased.

Converting the temperature radiography of the vapor cell in Figure 6 to the temperature error analysis diagram shown in Figure 7 in order to visualize the temperature uniformity of the vapor cell.

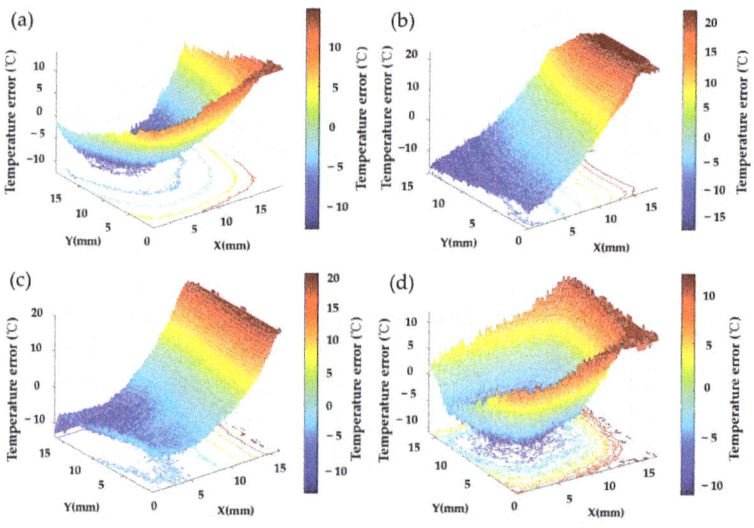

Figure 7. Temperature error profiles were measured for the vapor cell with different graphite films being added. (**a**) The vapor cell without graphite films; (**b**) the vapor cell with graphite film added to two surfaces; (**c**) the vapor cell with graphite films added to three surfaces; (**d**) the vapor cell with graphite films added to four surfaces.

It can be seen from Figure 7 that the graphite film can improve the thermal conductivity of the heating structure and balance the temperature distribution of the vapor cell. However, due to the heating chip of the incident window absorbing too much laser energy, the temperature of the incident window is higher. The reason why the two surfaces in the *Y*-axis direction of the vapor cell are not chosen to stick graphite films is to capture the temperature radiation pattern of the vapor cell.

Figures 6 and 7 show the influence of graphite film on the temperature uniformity of the vapor cell. Finally, we introduce the temperature difference coefficient to evaluate the temperature uniformity of the vapor cell.

The expression for the CVT is as follows:

$$CVT = \frac{f}{T_{aver}}, \tag{7}$$

where T_{aver} is the average temperature, and f is the unbiased variance of the temperature. T_{aver} denotes $\sum T_i / S$, where T_i is the temperature of each plane point, and S is the surface area; f is denoted by $\sqrt{\sum(T_i - T_{aver})^2 / S}$. According to Equation (7), the smaller the CVT is, the more uniform the temperature distribution is. The measured temperature radiography was converted into a temperature matrix to obtain the CVT values, and the different CVTs were obtained, as shown in Table 1.

Table 1. The CVT value of the vapor cell changes when graphite films of different sizes are applied to the outer wall of the vapor cell.

Size	CVT
0	0.1308
20 mm × 20 mm (two pieces)	0.1038
20 mm × 60 mm	0.0697
20 mm × 80 mm	0.0426

From Table 1, as the size of the graphite film increases, the calculated CVT of the vapor cell decreases, and the temperature uniformity of the vapor cell improves. Therefore, the temperature uniformity of the heating structure is greatly improved by adding graphite films.

5. Conclusions

In this paper, a vapor cell heating structure based on laser heating technology was constructed, and the spatial distribution of temperature was optimized via graphite films. By constructing the finite element model, the numerical analysis of the thermal transfer in the heating system shows that the uniformity of the temperature in the vapor cell is limited because the heat transfer is poor between the vapor cell and the heating chips. The laser power is constant, and the inner temperature CVT of the vapor cell decreased by 67% after adding graphite film on the four sides of the vapor cell, so the structure of graphite film can efficiently improve the temperature distribution homogeneity in the vapor cell. In the next step, we will study the action mechanism between the temperature homogeneity and the magnetometer output signal.

Author Contributions: Conceptualization, Y.L. and G.Z.; methodology, G.Z.; software, Y.L.; validation, Y.L., G.Z. and S.T.; formal analysis, Y.L.; investigation, G.Z.; resources, X.G.; data curation, Y.L.; writing—original draft preparation, X.L.; writing—review and editing, G.Z.; visualization, S.T.; supervision, S.T.; project administration, X.D.; funding acquisition, X.G. All authors have read and agreed to the published version of the manuscript.

Funding: This research was funded by the National Natural Science Foundation of China (NSFC), grant number 62005167.

Institutional Review Board Statement: Not applicable.

Informed Consent Statement: Not applicable.

Data Availability Statement: Not applicable.

Conflicts of Interest: The authors declare no conflict of interest.

References

1. Boto, E.; Meyer, S.S.; Shah, V.; Alem, O.; Knappe, S.; Kruger, P.; Fromhold, T.M.; Lim, M.; Glover, P.M.; Morris, P.G.; et al. A new generation of magnetoencephalography: Room temperature measurements using optically-pumped magnetometers. *Neuroimage* **2017**, *149*, 404–414. [CrossRef]
2. Boto, E.; Holmes, N.; Leggett, J.; Roberts, G.; Shah, V.; Meyer, S.S.; Munoz, L.D.; Mullinger, K.J.; Tierney, T.M.; Bestmann, S.; et al. Moving magnetoencephalography towards real-world applications with a wearable system. *Nature* **2018**, *555*, 657–661. [CrossRef] [PubMed]
3. Zhang, G.Y.; Huang, S.J.; Xu, F.X.; Hu, Z.H.; Lin, Q. Multi-channel spin exchange relaxation free magnetometer towards two-dimensional vector magnetoencephalography. *Opt. Express* **2019**, *27*, 597–607. [CrossRef]
4. Hill, R.M.; Boto, E.; Holmes, N.; Hartley, C.; Seedat, Z.A.; Leggett, J.; Roberts, G.; Shah, V.; Tierney, T.M.; Woolrich, M.W.; et al. A tool for functional brain imaging with lifespan compliance. *Nat. Commun.* **2019**, *10*, 4785. [CrossRef]
5. Wyllie, R.; Kauer, M.; Smetana, G.S.; Wakai, R.T.; Walker, T.G. Magnetocardiography with a modular spin-exchange relaxation-free atomic magnetometer array. *Phys. Med. Biol.* **2012**, *57*, 2619. [CrossRef]
6. Alem, O.; Sander, T.H.; Mhaskar, R.; LeBlanc, J.; Eswaran, H.; Steinhoff, U.; Okada, Y.; Kitching, J.; Trahms, L.; Knappe, S. Fetal magnetocardiography measurements with an array of microfabricated optically pumped magnetometers. *Phys. Med. Biol.* **2015**, *60*, 4797–4811. [CrossRef]
7. Jensen, K.; Skarsfeldt, M.A.; Staerkind, H.; Arnbak, J.; Balabas, M.V.; Olesen, S.P.; Bentzen, B.H.; Polzik, E.S. Magnetocardiography on an isolated animal heart with a room-temperature optically pumped magnetometer. *Sci. Rep.* **2018**, *8*, 16218. [CrossRef]
8. Perry, A.R.; Bulatowicz, M.D.; Larsen, M.; Walker, T.G.; Wyllie, R. All-optical intrinsic atomic gradiometer with sub-20 fT/cm/root Hz sensitivity in a 22 mu T earth-scale magnetic field. *Opt. Express* **2020**, *28*, 36696–36705. [CrossRef]
9. Dang, H.B.; Maloof, A.C.; Romalis, M.V. Ultrahigh sensitivity magnetic field and magnetization measurements with an atomic magnetometer. *Appl. Phys. Lett.* **2010**, *97*, 151110. [CrossRef]
10. Ito, Y.; Sato, D.; Kamada, K.; Kobayashi, T. Optimal densities of alkali metal atoms in an optically pumped K-Rb hybrid atomic magnetometer considering the spatial distribution of spin polarization. *Opt. Express* **2016**, *24*, 15391–15402. [CrossRef] [PubMed]
11. Huang, Q.; Xiang, K.; Huang, Z.J.; Sun, W.M.; Zhang, J.H. A non-magnetic temperature controlling system applied in atomic magnetometer. *Adv. Mat. Res.* **2013**, *760–762*, 896–900. [CrossRef]
12. Liu, G.B.; Li, X.F.; Sun, X.P.; Feng, J.W.; Ye, C.H.; Zhou, X. Ultralow field NMR spectrometer with an atomic magnetometer near room temperature. *J. Magn. Reson.* **2013**, *237*, 158–163. [CrossRef] [PubMed]
13. Jiang, Q.Y.; Luo, H.; Zhan, X.; Zhang, Y.; Yang, K.Y.; Wang, Z.G. Avoiding the impact of the heater-induced longitudinal field on atomic magnetometers. *J. Appl. Phys.* **2018**, *124*, 244501. [CrossRef]
14. Liang, X.Y.; Liu, Z.C.; Die, H.; Wu, W.F.; Jia, Y.C.; Fang, J.C. MEMS Non-Magnetic Electric Heating Chip for Spin-Exchange-Relaxation-Free (SERF) Magnetometer. *IEEE Access* **2019**, *7*, 88461–88471.
15. Lu, J.X.; Wang, J.; Yang, K.; Zhao, J.P.; Quan, W.; Han, B.C.; Ding, M. In-Situ Measurement of Electrical-Heating-Induced Magnetic Field for an Atomic Magnetometer. *Sensors* **2020**, *20*, 1826. [CrossRef]
16. Schwindt, P.D.D.; Lindseth, B.; Knappe, S.; Shah, V.; Kitching, J.; Liew, L.A. Chip-scale atomic magnetometer with improved sensitivity by use of the M-x technique. *Appl. Phys. Lett.* **2007**, *90*, 081102. [CrossRef]
17. Allred, J.C.; Lyman, R.N.; Kornack, T.W.; Romalis, M.V. High-sensitivity atomic magnetometer unaffected by spin-exchange relaxation. *Phys. Rev. Lett.* **2002**, *89*, 130801. [CrossRef]
18. Kominis, I.K.; Kornack, T.W.; Allred, J.C.; Romalis, M.V. A subfemtotesla multichannel atomic magnetometer. *Nature* **2003**, *422*, 596–599. [CrossRef] [PubMed]
19. Preusser, J.; Gerginov, V.; Knappe, S.; Kitching, J. A microfabricated photonic magnetometer. In Proceedings of the 2008 IEEE Sensors, Lecce, Italy, 26–29 October 2008; pp. 344–346.
20. Alem, O.; Mhaskar, R.; Jimenez-Martinez, R.; Sheng, D.; LeBlanc, J.; Trahms, L.; Sander, T.; Kitching, J.; Knappe, S. Magnetic field imaging with microfabricated optically-pumped magnetometers. *Opt. Express* **2017**, *25*, 7849–7858. [CrossRef]
21. Oelsner, G.; Schultze, V.; Ijsselsteijn, R.; Wittkamper, F.; Stolz, R. Sources of heading errors in optically pumped magnetometers operated in the Earth's magnetic field. *Phys. Rev. A* **2019**, *99*, 013420. [CrossRef]
22. Savukov, I.; Boshier, M.G. A High-Sensitivity Tunable Two-Beam Fiber-Coupled High-Density Magnetometer with Laser Heating. *Sensors* **2016**, *16*, 1691. [CrossRef] [PubMed]
23. Preusser, J.; Knappe, S.; Kitching, J.; Gerginov, V. A microfabricated photonic magnetometer. In Proceedings of the 2009 Joint Meeting of the European Frequency and Time Forum and the IEEE International Frequency Control Symposium, VOLS 1 and 2, Besancon, France, 20–24 April 2009; pp. 1180–1182.
24. Mhaskar, R.; Knappe, S.; Kitching, J. A low-power, high-sensitivity micromachined optical magnetometer. *Appl. Phys. Lett.* **2012**, *101*, 241105. [CrossRef]
25. Sheng, D.; Perry, A.R.; Krzyzewski, S.P.; Geller, S.; Kitching, J.; Knappe, S. A microfabricated optically-pumped magnetic gradiometer. *Appl. Phys. Lett.* **2017**, *110*, 031106. [CrossRef]
26. Puri, P.; Jordan, P.M. Wave structure in Stokes' second problem for a dipolar fluid with nonclassical heat conduction. *Acta Mech.* **1999**, *133*, 145–160. [CrossRef]
27. Nika, D.L.; Balandin, A.A. Phonons and thermal transport in graphene and graphene-based materials. *Rep. Prog. Phys.* **2017**, *80*, 036502. [CrossRef]

28. Cao, H.Y.; Guo, Z.X.; Xiang, H.J.; Gong, X.G. Layer and size dependence of thermal conductivity in multilayer graphene nanoribbons. *Phys. Lett. A* **2012**, *376*, 525–528. [CrossRef]
29. Gill-Comeau, M.; Lewis, L.J. Heat conductivity in graphene and related materials: A time-domain modal analysis. *Phys. Rev. B* **2015**, *92*, 195404. [CrossRef]
30. Vora, H.D.; Santhanakrishnan, S.; Harimkar, S.P.; Boetcher, S.K.S.; Dahotre, N.B. One-dimensional multipulse laser machining of structural alumina: Evolution of surface topography. *Int. J. Adv. Manuf. Tech.* **2013**, *68*, 69–83. [CrossRef]

photonics

MDPI

Article

Far-Detuning Laser Frequency Disturbance Suppression for Atomic Sensor Based on Intrinsic Fiber Fabry–Pérot Cavity

Guanghui Li [1,2], Lihong Duan [1,2,*], Xinxiu Zhou [1,2,3,4] and Wei Quan [1,2,*]

1 School of Instrumentation and Optoelectronic Engineering, Beihang University, Beijing 100191, China; guanghh@buaa.edu.cn (G.L.); zhouxinxiu@buaa.edu.cn (X.Z.)
2 Institute of Large-scale Scientific Facility and Centre for Zero Magnetic Field Science, Beihang University, Beijing 100191, China
3 National Institute of Extremely-Weak Magnetic Field Infrastructure, Hangzhou 310028, China
4 College of Metrology Measurement and Instrument, China Jiliang University, Hangzhou 314423, China
* Correspondence: duanlihong@buaa.edu.cn (L.D.); quanwei@buaa.edu.cn (W.Q.)

Abstract: The method of laser far-detuned frequency locking is proposed based on a fiber Fabry–Perot cavity which transfers the ultra-stable atomic reference frequency stability to the target laser utilized for atomic sensors. The control transfer function of the closed-loop system is established to elucidate the process of perturbation suppression. It is illustrated that this method is robust against the disturbance to the laser and cavity by controlling the cavity with different parameters. After the long-term experimental test, the stability of the laser frequency locked on the fiber cavity achieves an Allan deviation of 9.9×10^{-11} and the detuning of the nearest atomic frequency resonance point is more than 200 GHz. Its stability and detuning value exceed previous reports.

Keywords: laser frequency; frequency detuning and stabilization; fiber Fabry–Perot cavity; transfer control loop

1. Introduction

The latest generation of ultra-high-sensitivity quantum sensors, exemplified by atomic magnetometers and gyroscopes, are pivotal for measuring biological and Earth magnetic fields, as well as inertial information [1–3]. These devices typically necessitate a pump laser to polarize the atoms and a probe laser to detect the signal [4]. The pump laser operates at the atomic hyperfine resonance frequency. Deviations from this resonance point can result in light shifts and magnetic noise [5,6]. Conversely, the probe laser operates at the far-detuned frequency relative to the resonance point, at a detuning range from 100 to 200 GHz, to confirm the optimal scale coefficient. Frequency fluctuations of the pump laser and probe laser decrease both the long-term stability and sensitivity [7].

In previous studies on spin-exchange relaxation-free (SERF) atomic sensors, the pump laser typically achieves long-term frequency locking through techniques such as saturated absorption frequency stabilization [8], circular dichroism frequency stabilization [9], and polarization spectrum frequency stabilization [10]. But the probe laser still lacks an effective detuning frequency stabilization method because it is challenging to simultaneously achieve large detuning, long-term stability, and prototyping with the exiting active frequency control techniques. Recently, WeiYao attempted an in situ frequency detuning stabilization method in a SERF co-magnetometer, while the frequency stability after locking was still poor, and the frequency drift reached more than 100 MHz/5 h [7]. Within the atomic clock domain, detuning the laser frequency to the magic wavelength often involves using an optical frequency comb [11]. Despite the comb's ability to offer broad repetition frequencies for extensive detuning and locking, its long-term stability relies on the stability of a hydrogen maser clock [12]. Alternatively, an ultra-stable cavity constructed from materials with extremely low thermal expansion coefficients can also generate a comb-shaped resonant frequency for frequency detuning and locking, but it requires operation in a vacuum and temperature-controlled environment to mitigate frequency drift [13,14]. Consequently, a frequency detuning locking method for probe lasers that can ensure miniaturization and high stability at the same time is urgently needed.

By reusing the pump laser frequency of the atomic sensor as a reference and compensating the drift and jitter of the cavity, the frequency stability of the resonant reference is transferred to the far-detuned target laser. To analyze the intriguing phenomenon why the transferred frequency stability could exceed the reference frequency, we established the transfer function of the control system and studied the main factors affecting the transferred laser frequency stability, including discriminator noise, open-loop gain, and the discriminator coefficient determined by the cavity parameters. The frequency stability was evaluated using the noise power spectral density of the error signal and the Allan variance of the wavelength data after locking. This method facilitates miniaturization and sustains high-precision frequency control.

2. Theory

2.1. Principle of Transferred FFP Frequency Locking

The intrinsic optical fiber Fabry–Perot (FFP) cavity is made by coating parallel end faces on both ends of a single-mode optical fiber, allowing for cavity mode coupling without the need for complex mirror adjustments [15]. According to the Jones matrix of the cavity, the reflected light field after resonating in the optical FFP cavity is expressed as

$$\begin{bmatrix} E_{\text{out}\,x} \\ E_{\text{out}\,y} \end{bmatrix} = \begin{bmatrix} M_x & 0 \\ 0 & M_y \end{bmatrix} \cdot \begin{bmatrix} E_{\text{inx}} \\ E_{\text{iny}} \end{bmatrix} \tag{1}$$

where $\begin{bmatrix} E_{\text{outx}} \\ E_{\text{outy}} \end{bmatrix}$ is the Jones vector of the reflected light field, $\begin{bmatrix} E_{inx} \\ E_{iny} \end{bmatrix}$ is the Jones vector of the input light field. $M_x = R - R(1-R^2)e^{j2\delta_x} \cdot \sum_{n=0}^{n=h}\left(R^2 e^{j2\delta_x}\right)^n$, and $M_y = R - R(1-R^2)e^{j2\delta_y} \cdot \sum_{n=0}^{n=h}\left(R^2 e^{j2\delta_y}\right)^n$. R is the coating reflectivity ($0 < R < 1$). δ_x and δ_y are the phase produced by the reflection light along the x polarization axis and the y polarization axis of the fiber FP cavity, respectively. So when the input light is a beam of linearly polarized light, the azimuth angle of which is determined by the polarization axis of the polarization-maintaining fiber circulator (PM-OPC) ($\begin{bmatrix} E_{inx} \\ E_{iny} \end{bmatrix} = \begin{bmatrix} \cos\alpha \\ \sin\alpha \end{bmatrix}$), the intensity of the output reflection spectrum is stable. Based on $I_{R_{x,y}} = \left|E_{\text{out}\,x,y}\right|$, the intensity is calculated as

$$I_{R_x} = \frac{\cos^2\alpha R^2}{1+\frac{(R-1)^2}{4R(\sin\delta_x)^2}}, I_{R_y} = \frac{\sin^2\alpha R^2}{1+\frac{(R-1)^2}{4R(\sin\delta_x)^2}} \tag{2}$$

From Equation (2), it can be seen that the reflection spectrum is Lorentz line-type, and when the phase satisfies $\delta_{x,y} = k\pi$, the spectral intensity reaches the maximum. According to the relationship between the phase and the laser frequency in the cavity, $\delta_{x,y} = \frac{2\pi n_{x,y} L v_{x,y}}{c}$, where $n_{x,y}$ is the refractive index of light transmitted in the optical fiber. L is the fiber length and $v_{x,y}$ is the laser frequency.

The corresponding resonant frequencies of the two orthogonal polarization modes at the maximum intensity are calculated as

$$v_{q_{x,y}} = \frac{qc}{2n_{x,y}L} \tag{3}$$

where q is an integer that represents the q-th mode.

Due to the birefringence of the fiber FP cavity, the fiber FP cavity can simultaneously generate two sets of orthogonal resonance peaks. And its resonance frequency is not only related to the cavity length, but also related to the refractive index of the two polarization axes. So the single FFP used as the frequency reference is much more sensitive than the traditional quartz FP cavity. On the other hand, this also means that the FFP cavity is much easier to control. Let $\tau = \frac{nL}{c}$, so the resonance frequency is expressed as

$$v = \frac{q}{2\tau} \tag{4}$$

The delay τ of the cavity can always be adjusted to resonant with the frequency-stabilized laser by adjusting the cavity temperature or applying elastic stress to the cavity. Considering the thermal and elastic effects of the optical fiber, $\Delta\tau \approx K \cdot \Delta T + J \cdot L \cdot \varepsilon$ [16], where K is the thermal-optic coefficient, J is a constant related to the elastic-optic coefficient, ΔT is the temperature difference of the cavity, and ε represents the strain applied to the cavity. As the response bandwidth of PZT (dozens KHz) is much higher than that of TEC (several Hz), so the thermal drift of the cavity with the ambient temperature is compensated easily by elastic control.

2.2. Locked Frequency Stability Analysis

In this paper, the demodulated error signal of the cavity and lasers are all dispersive lines. The pump laser frequency is locked on the saturated absorbed peak of the Rb atomic D1 line. Then, the FFP cavity phase delay (τ) is locked on the reflected spectrum of the pump laser resonating within the cavity. Lastly, the probe laser frequency is locked on the resonant spectrum of the FFP cavity's phase delay. The reference of the saturated absorption line and the phase delay of the FFP cavity are converted to error signal zeros by the discriminator. The entire control process is shown in Figure 1. The residual frequency noise of the pump laser (output from loop 1) is converted to the input noise of loop 2. Identically, the residual frequency noise of loop 2 is added to the input noise of loop 3.

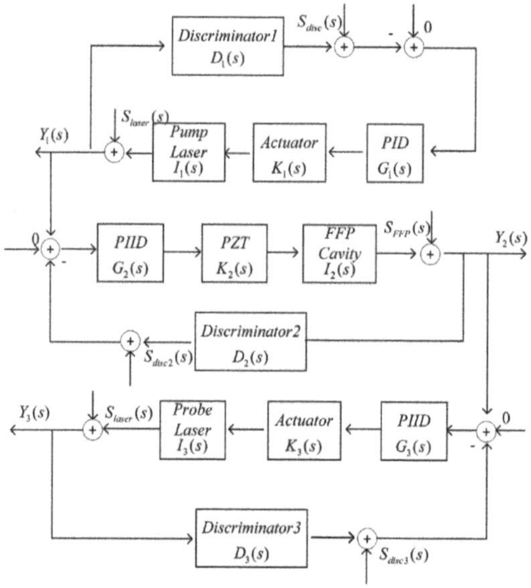

Figure 1. The control block diagram of the transferring locking system.

In the block diagram, the main disturbance is introduced from the discriminator and the environment, and the control noise and actuator noise are disregarded. The discriminator noise [17] is expressed as

$$S_{\text{disc}}(f) = kQ\sqrt{NEP^2 + 2\hbar v P_{\text{det}}/\eta_d} \tag{5}$$

where k is the conversion loss of the frequency discriminator. Q is the responsivity constant determined by the photodetector and the pre-amplifier. η_d is a constant representing the quantum efficiency of the PD. Therefore, the $S_{\text{disc}}(f)$ is mainly affected by the intensity noise in the frequency domain of the input spectrum, and there is no correlation between the frequency discrimination noise in the three loops. $S_{\text{disc}}(s)$ is the expression of the Laplace domain. The frequency noise of a free-running laser subject to environmental disturbances in the s-domain is expressed as $S_{\text{laser}}(s)$. According to the Wiener–Khintchine theorem, the modulation noise broadens the line-width of the laser [18], but has little effect on long-term stability. Due to the bandwidth limitations of the PID closed-loop control system, the line-width noise caused by modulation is not considered here. Therefore, the primary term of laser frequency noise is the slow drift which limits the long-term stability. The SMF-FP noise in the s-plane is expressed as $S_{FFP}(s)$, which is mainly composed of $1/f$ noise and random polarization fluctuation.

According to Figure 1, the output noise of the pump laser, the output noise of the FFP cavity, and the output noise of the probe laser are expressed as

$$Y_1 = \frac{A_1}{1 + A_1 B_1} S_{\text{disc1}}(s) + \frac{S_{\text{laser}}(s)}{1 + A_1 B_1} \tag{6}$$

$$Y_2 = \frac{A_2}{1 + A_2 B_2} (S_{\text{disc2}}(s) + Y_1(s)) + \frac{S_{FFP}(s)}{1 + A_2 B_2} \tag{7}$$

$$Y_3 = \frac{A_3}{1 + A_3 B_3} (S_{\text{disc3}}(s) + Y_2(s)) + \frac{S_{\text{laser}}(s)}{1 + A_3 B_3} \tag{8}$$

where $A_1 = I_1 K_1 G_1$, $B_1 = D_1$; $A_2 = I_2 K_2 G_2$, $B_2 = D_2$; and $A_3 = I_3 K_3 G_3$, $B_3 = D_3$. Both I and K are constants determined by the device. When $A_1 \gg 1$, $Y_{\text{loop 1}} \approx \frac{S_{\text{disc1}}(s)}{B_1}$. After the loop becomes a closed loop, the low-frequency slow drift of the laser frequency is greatly suppressed, and the remaining frequency noise is approximately high-frequency white noise.

As the FFP cavity locking loop adopts phase modulation with 100 MHz, the closed-loop control bandwidth of loop 2 reaches the MHz level.The PIID controller in loop 2 is employed to increase the open-loop gain. As indicated by Equation (7), when the open-loop gain is sufficiently high, the slow drift ($S_{FFP}(s)$) of the cavity is significantly suppressed, leaving only the extremely weak frequency discrimination noise $S_{\text{disc2}}(s)$. Then, by substituting the simplified $Y_1(s)$ into Equation (4), $Y_2(s)$ is simplified as

$$Y_2(s) \approx \frac{1}{B_2} S_{\text{disc2}}(s) + \frac{1}{B_2 B_1} S_{\text{disc1}}(s) \tag{9}$$

The loop 3 bandwidth is also limited by the cut-off frequency of the discriminator, and the simplified Equation (8) is expressed as

$$Y_3(s) \approx \frac{S_{\text{disc3}}(s)}{B_3} + \frac{S_{\text{disc2}}(s)}{B_2 B_3} + \frac{S_{\text{disc1}}(s)}{B_1 B_2 B_3} \tag{10}$$

According to Equation (10), the transferred probe laser frequency noise is mainly composed of discrimination noise. The low drift of the probe laser frequency is significantly suppressed under the influence of three closed-loop loops. It is seen that the influence of the three kinds of frequency discrimination noise is $S_{\text{disc3}}(s) > S_{\text{disc2}}(s) > S_{\text{disc1}}(s)$. And by optimizing the coefficients of the frequency discriminator of the three loops, the output noise of the probe laser can be further reduced. Notably, when large environmental disturbance ($S_{FFP}(s)$) is introduced to the cavity in loop 2, it can easily cause the integral saturation of the cavity and the whole system is no longer locked. Therefore, it is still necessary to add passive temperature control to the cavity to reduce the noise transmitted to the cavity in loop 3, which ensures the long-term stability of the entire system.

3. Experimental Setup

The experiment setup of the transferred frequency stabilization system is illustrated in Figure 2. The homemade 780 nm semiconductor laser (of which the diode style is EYP-DFB-0780-00080-150) is divided into two beams by the polarizing beam splitter (PBS), and one part of the light beam is transmitted in the Rubidium cell (Rb Cell) to lock the pump laser frequency on the saturated absorption peak. The error signal is obtained by the demodulation method and controlled with a Proportional-Integral-Dfferential (P-I-D) controller. The control loop is shown as Loop 1 in Figure 1. The other part of the beam light is coupled into the fiber splitter to pump the atoms (or monitor the stabilized frequency) and to stabilize the tunable FFP cavity.

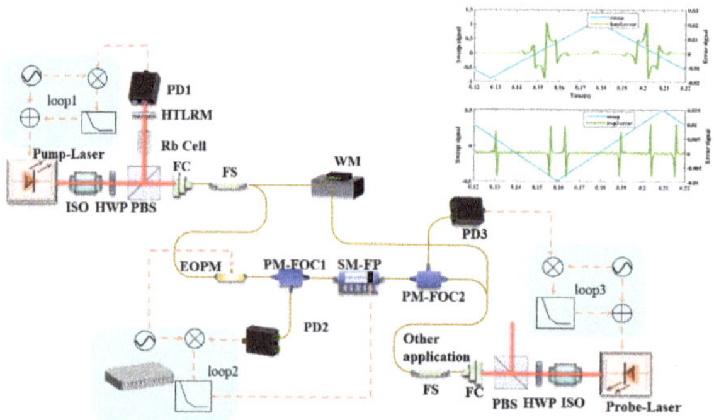

Figure 2. Experiment setupof the transferred frequency locking system. ISO: Isolator; HWP: Half-wave plane; PBS: Polarizing beam splitter; HTLRM: High-transmittance low reflective mirror; FC: Fiber coupler; FS: Fiber splitter; EOPM: Electro-optic phase modulator; PM-FOC: Polarization-maintaining fiber optic circulator; PD: Photo-detector; WM: Wavelength meter.

The tunable transferred fiber cavity is composed of an FFP, two thermoelectric cooler (TEC) plates, two negative thermistors, and a piezoelectric transducer (PZT) actuator. The FFP cavity is made of a single-mode fiber plating high-reflection coating (>98%), the fineness of which is about 125 and the Full Width at Half Maximum (FWHM) is about 8.25 MHz [19]. Its resonance characteristics have been reported in another study.

As directly modulating the cavity's PZT causes the vibration noise of the FFP, an electro-optic phase modulator (EOPM) is used for modulating the light phase to stabilize the FFP cavity. The PDH technique is used to obtain the error signal and the proportional-integral-integral-differential controller (P-I-I-D controller) at loop 2, seen in Figure 1. This is applied to control the cavity stability. The modulated light transmits the polarization-maintaining fiber optic circulator (PM-FOC) from port 1 to port 2, and then inputs it into the FFP cavity. The reflected spectrum is output from port 3 of PM-FOC-1. Benefiting from PM-FOC's polarization-maintaining and the nonreciprocal properties, the input light remains linearly polarized and the reflected light does not interfere with the input light.

The 795 nm laser used in the experiment is a distributed grating feedback semiconductor laser (EYP-DFB-0795-00080), which has the advantage of a narrow line-width of single longitudinal mode. The light emitted by the laser is divided into two beams of light with different power. One is used as the probe light in the atomic sensors, and the other beam is coupled into the fiber splitter for the purpose of frequency detuned locking. Considering the simplification of the optical pathway, the frequency modulation of the 795 nm laser is achieved through current modulation, with a bandwidth capability of 4 MHz. The modulated light is transmitted to the fiber FP cavity through PM-FOC-2. Its reflected resonant spectrum is received by a photo detector. Similarly, demodulating the modulated spectrum achieves the error signal and controls the frequency in a closed loop denoted as loop 3 in Figure 1.

It is noteworthy that the photo-diode bandwidth within the three loops depicted in Figure 1 is contingent upon their respective modulation bandwidths. As the modulation frequency ranges in loop 1, loop 2, and loop 3 are 10 KHz to 50 KHz, 10 MHz to 100 MHz, and 200 KHz to 1.5 MHz, respectively, the PD bandwidths of loop 1, loop 2, and loop 3 are 100 KHz, 350 MHz, and 15 MHz, respectively.

4. Experimental Results

4.1. Parameter Optimization

The frequency discrimination slope (D-slope) of the three loops' discriminator under various conditions is tested as seen in Figure 3. Limited by the line-width of the saturated absorption spectrum, B_1 is approximately a constant whose impact on the output is also minimal. B_2 and B_3 are affected by modulation depth, modulation frequency, and the line-width of fiber FP cavity. For phase modulation in loop 2, the D-slope increases with modulation depth and the fineness of the FFP cavity, independent of modulation frequency. However, for frequency modulation in loop 3, achieved through laser current modulation, the D-slope exhibits more complex variations. Experimental results indicate that beyond a certain modulation depth, the D-slope decreases, possibly due to the increased laser line-width from current modulation. The D-slopes corresponding to different parameters are experimentally selected to optimize the control performance. Due to the low sample rate of the wavelength meter, the frequency stability in the loop is evaluated by calculating the noise power spectral density (PSD) in the actual experiment.

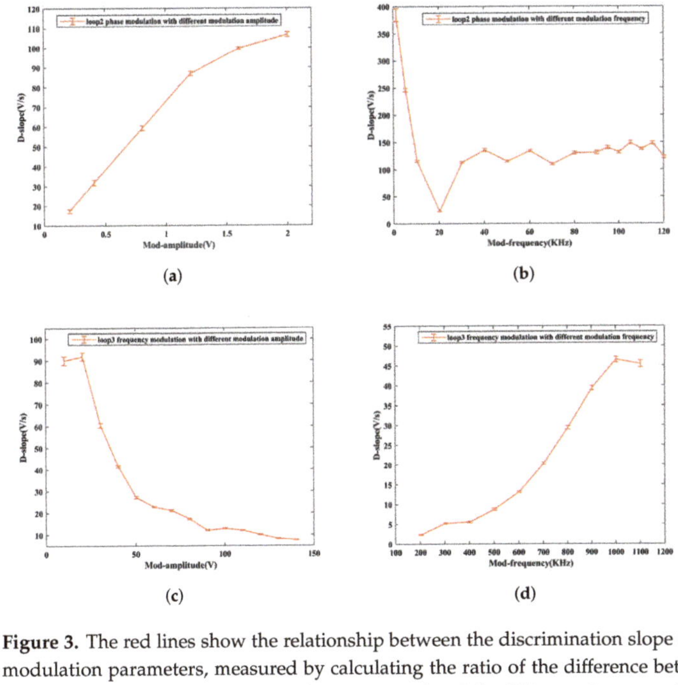

Figure 3. The red lines show the relationship between the discrimination slope (D-slope) and the modulation parameters, measured by calculating the ratio of the difference between the voltage between the two extreme values of the error signal and the difference between the abscissa sweep times. The error bar is obtained by three repeated independent measurements. (**a**) The D-slope vs. the phase modulation amplitude of loop 2; (**b**) the D-slope vs. the phase modulation frequency of loop 2; (**c**) the D-slope vs. the frequency modulation amplitude of loop 3; (**d**) the D-slope vs. the frequency modulation frequency of loop 3.

The error signals of the probe laser in both open loop and closed loop states are collected to calculate the system PSD that are depicted in Figure 4. It is shown that the frequency noise in the three loop bandwidths is greatly suppressed compared to the open-loop state. And compared to the single integrator control method, the double-integral closed-loop controller has a better disturbance suppression effect.

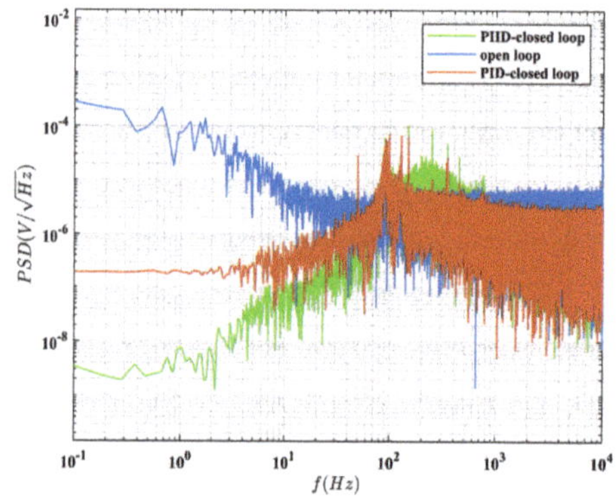

Figure 4. The blue, orangeand green lines are the power spectral densitie (PSD) lines of the probe laser's error signals under the open loop, the closed loop using the P-I-D control method, and the closed loop using the P-I-I-D control method, respectively.

According to Equations (9) and (10), when the D-slope value of loop 2 (B_2) is increased, the PSD of loop 2 will decrease and the PSD of loop 3 will not change significantly as the frequency noise of loop 3 is mainly determined by S_{disc_3}. Figure 5 shows the tested PSD of loop 2 and loop 3 by varying the modulation amplitudes of loop 2. The results are consistent with the analysis (the signal sampling rate for calculating PSD is 500 KHz). Since the bandwidth of loop 2 can reach MHz, the PSD of loop 2 shown in Figure 5a has only one corner frequency, f_1 (f_1 = 120 Hz), which is determined by the cut-off frequency of the integrator of loop 2.

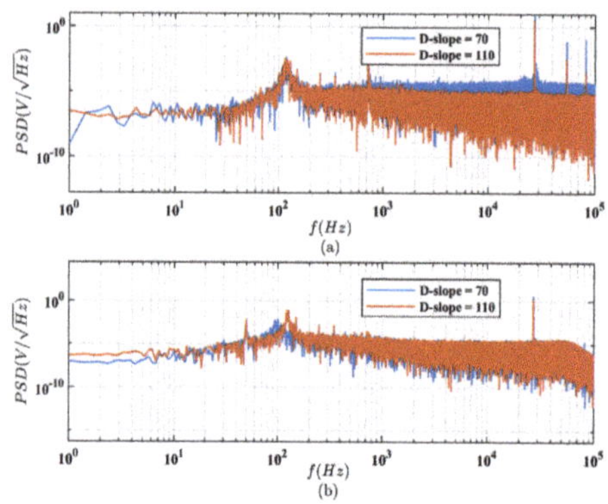

Figure 5. The loop 2 PSD line of the cavity error signal and the loop 3 PSD line of the probe laser error signal measured by varying the modulation amplitude in loop 2. (**a**) The PSD of loop 2. The blue and orange lines are the PSD lines measured when the modulation amplitudes of loop 2 are equal to 1 V and 2 V, respectively, and D-slope = 70 V/s and D-slope = 110 V/s. (**b**) The PSD of loop 3. The blue and orange lines are the PSD when the modulation amplitudes of loop 2 are equal to 1 V and 2 V, respectively, and D-slope = 70 V/s and D-slope = 110 V/s.

Similarly, the D-slope of loop 3 is adjusted by varying the current modulation frequency. As shown in Figure 6a, the B3 at 1000 KHz of loop 3 is higher than it is at 500 KHz. And the closed-loop frequency noise corresponding to the modulation frequency of 1000 KHz is lower than it is in the condition of 500 KHz, which is all tested above the integral crossover frequency. It is shown in Figure 6b that the frequency noise declines when increasing the open-loop gain of the controller. By appropriately increasing the open-loop gain, a better noise suppression effect can be obtained. The results are consistent with the analysis.

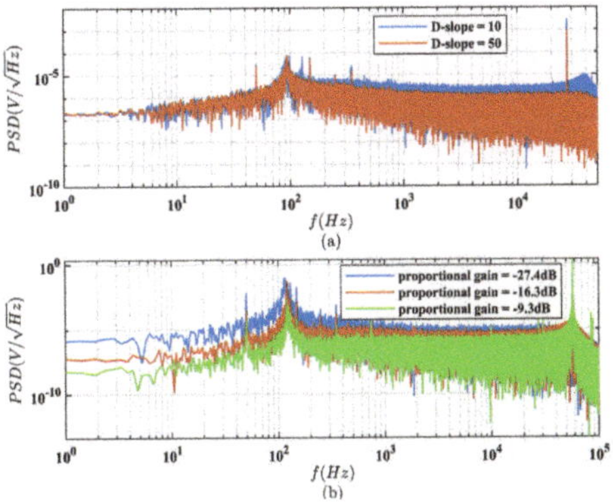

Figure 6. The loop 3 PSD lines of the probe laser error signal measured by varying the modulation frequency and the proportional gain in loop 3. (**a**) The blue and orange lines are the PSD lines when the modulation frequencies of loop 2 are equal to 500 KHz and 1000 KHz, respectively, and D-slope = 10 V/s and D-slope = 50 V/s. (**b**) The blue, orange, and green lines are the PSD lines when the proportional gains of the control loop equal −27.4 dB, −16.3 dB, and −9.3 dB, respectively.

This experiment also elucidates why the final frequency stability of loop 3 has the potential to surpass that of loop 1. Although the latter loop uses the previous loop as the locking reference, the control performance in each loop is mostly influenced by the D-slope and the control parameters of its own loop. The D-slope for saturated absorption frequency stabilization in loop 1 is much lower than that of the Pound-Drever-Hall (PDH) frequency stabilization system in loop 3, and thuse the transfer frequency stabilization method based on the FFP cavity achieves superior stability.

4.2. *Pump and Probe Laser Locking Results*

After optimization, the modulation parameters of loop 2 are set as $f_{\text{mod}} = 60$ MHz and $A_{\text{mod}} = 2$ V, and the modulation parameters of loop 3 are set as $f_{\text{mod}} = 1.1$ MHz and $A_{\text{mod}} = 30$ mV. The referenced laser (pump laser) and the transferred laser (probe laser) frequency are sampled by a wavelength meter (Highfiness, WS7-60) for 2 days to measure the long-term stability. Figure 7 shows the wavelength fluctuation and drift of the pump laser, probe laser, and free-running probe laser. The probe laser wavelength fluctuation after locking on the transferred FFP cavity is only 2.3 MHz within an hour, while it reaches 19.9 MHz in the free-running state. Moreover, as the measure time increases, the difference in wavelength drift is more significant in the free-running and locked states. The pump laser wavelength locked with the saturation absorb line, the probe laser wavelength locked with the transferred FFP cavity, and the free-running probe laser wavelength drift are 6.4 MHz, 5.2 MHz, and 66.7 MHz, respectively, in 48 h. For a more accurate analysis, Figure 8 illustrates the Allan deviation based on their wavelength data. The results indicate that the Allan deviation of the laser locked on the SAS line and that locked on the transferred FFP cavity is 7.8^{-11} and 9.9^{-11}, respectively, at the integration time of 100 s, which is two orders of magnitude better than that of the free-running laser. The locking results are compared with other methods shown in Table 1.

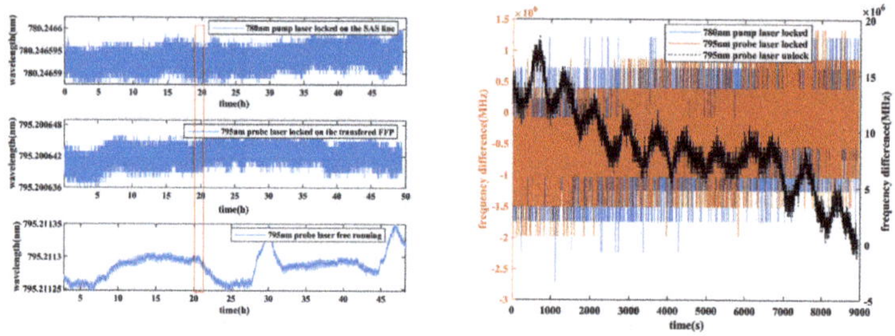

(a) (b)

Figure 7. The pump laser and probe laser wavelength drift trend after 48 h measured with the wavelength meter (**b**) The laser frequency variation details during one hour, which are marked with the red box in (**a**).

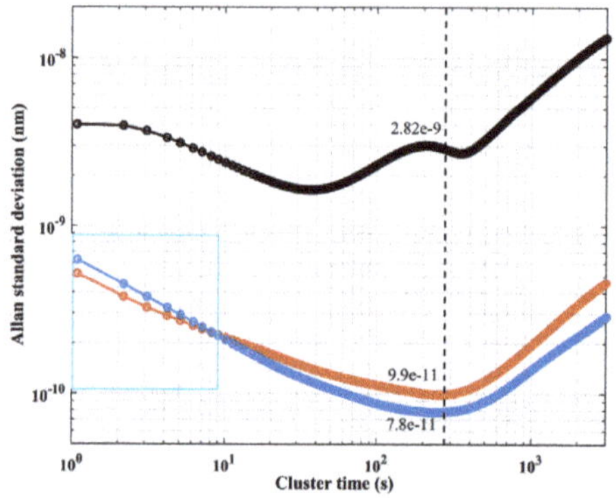

Figure 8. The black line is the Allan variance of the probe laser in the free-running state. The blue line is the Allan variance of the pump laser locking on the saturated absorption line. The orange line is the Allan variance of the probe laser locking on the transferred FFP cavity.

Table 1. Comparison of laser long-term stabilization performance locking by different methods

Locking Method	Locking Duration	Locking Performance
Presented in this article	48 h	Drift 6.4 MHz
Locked on a fiber Fabry Perot cavity [20]	35 h	Drift 10 MHz
Locked on open-loop PZT—controlled Fabry Perot cavity [21]	41 h	Drift 11 MHz
Injection—locked to a fiber ring resonator [22]	15 min	Drift less than 15 MHz

5. Discussion

As Figure 8 indicates, when the integration time is less than 1 s marked with the blue box, the Allan deviation of the 795 nm probe laser is lower than the 780 nm referenced pump laser in this method. Due to the influence of temperature fluctuation caused by air conditioning on the cavity, $S_{FFP}(s)$ in loop 2 is superimposed on loop 3, which reduces the long-term frequency stability of the 795 nm probe laser. Through the analysis of Equations (7) and (8), increasing the frequency discrimination slope of loop 2 and loop 3 can distinctly suppress this negative effect.

Consequently, the frequency stability of the laser controlled in this method can be further enhanced by increasing the D-slope, such as by improving the coating reflectivity of the fiber FP cavity or optimizing the cavity length. Otherwise, spectral broadening caused by internal current modulation makes the D-slope decrease. If the error signals of loop 1 and loop 3 are obtained by modulating the laser phase using an external phase modulator, the laser frequency noise will be further reduced.

6. Conclusions

In this study, the laser frequency transferred locking from 780.2 nm to 795.2 nm wasrealized by controlling a coated fiber FP cavity. By analyzing the transfer function of the control loop, the factors constraining the transferred laser frequency stability were investigated and the strategies enhancing its locking performance were explored. Although the laser has no stable natural frequency reference at the detuning frequency point, it has the potential to surpass the laser frequency stability locked on the atomic transition line by designing a high-fineness transferred FFP. The FFP cavity is employed as the detuned frequency reference and exhibits excellent long-term performance for the first time to our knowledge.

Author Contributions: Conceptualization, G.L. and L.D.; Data curation, G.L.; Formal analysis, G.L.; Funding acquisition, W.Q. and X.Z.; Investigation, G.L.; Methodology, G.L. and L.D.; Project administration, W.Q. and X.Z.; Supervision, X.Z.; Validation, X.Z.; Visualization, L.D.; Writing—original draft, G.L.; Writing—review and editing, G.L. and L.D. All authors have read and agreed to the published version of the manuscript.

Funding: This work was supported by the National Science Fund for Distinguished Young Scholars (61925301); the National Natural Science Foundation of China (62122009, 62003024); and the Innovation Program for Quantum Science and Technology (2021ZD0300501).

Institutional Review Board Statement: Not applicable.

Informed Consent Statement: Not applicable.

Data Availability Statement: The data presented in this study are available on reasonable request from the corresponding author.

Conflicts of Interest: The authors declare no conflicts of interest.

References

1. Tian, M.; Quan, W.; Jiang, L.; Liu, J.; Lu, Z. Single-beam NMOR atomic magnetometer based on a fiberized EOM. *Opt. Lett.* **2023**, *48*, 3075–3078. [CrossRef] [PubMed]
2. Yuan, Z.; Lin, S.; Liu, Y.; Tang, J.; Zhai, Y. Gradient phase and amplitude errors in atomic magnetic gradiometers for bio-magnetic imaging systems. *iSicence* **2024**, *27*, 109250. [CrossRef] [PubMed]
3. Liu, X.; Bi, W.; Li, Y.; Hu, Y.; Chang, M.; Zhang, X. A review of fiber-coupled atomic magnetometer. *Microw. Opt. Technol. Lett. Opt. Technol.* **2023**, *65*, 1516–1524. [CrossRef]
4. Liu, S.; Wang, R.; Yuan, L.; Wu, J.; Yuan, Q.; Zhu, J.; Fan, W.; Wang, Z.; Du, P. Transverse light-shift in a spin-exchange relaxation-free co-magnetometer: Measurement, decoupling, and suppression. *Opt. Express* **2022**, *30*, 15310–15326. [CrossRef]
5. Wu, Z.; Pang, H.; Wang, Z.; Fan, W.; Liu, F.; Liu, Y.; Quan, W. A New Design Method for a Triaxial Magnetic Field Gradient Compensation System Based on Ferromagnetic Boundary. *IEEE Trans. Ind. Electron.* **2024**, *71*, 13361–13371. [CrossRef]
6. Jia, Y.; Liu, Z.; Ding, M.; Chai, Z.; Liang, X.; Wu, W. The effect of tensor light shift on residual magnetic field compensation in a nuclear spin co-magnetometer. *Appl. Phys. Lett.* **2020**, *116*, 142405. [CrossRef]
7. Wei, Y.; Liu, S.; Liu, F.; Fan, W.; Zhai, Y.; Quan, W. The optimization and stabilization of detuned probe light frequency to suppress electron spin polarization error in SERF co-magnetometer. *Sens. Actuators A Phys.* **2023**, *359*, 114467. [CrossRef]
8. Seltzer, J.S. Developments in Alkali-Metal Atomic Magnetometry. Ph.D. Thesis, Princeton University, Princeton, NJ, USA, 2008.
9. Talker, E.; Arora, P.; Zektzer, R.; Sebbag, Y.; Dikopltsev, M.; Levy, U. Light-Induced Atomic Desorption in Microfabricated Vapor Cells for Demonstrating Quantum Optical Applications. *Phys. Rev. Appl.* **2021**, *15*, L051001. [CrossRef]
10. Qi, H.; Yang, B.; Zhao, H.; Ling, X.; Deng, J.; Cheng, H. Narrow Linewidth Laser Frequency Stabilization System for Integrating Sphere Cold Atom Clock. *Laser Optoelectron. Prog.* **2023**, *60*, 1514008. [CrossRef]
11. Duan, Y.; Huang, Y.; Li, Y.; Wang, Y.; Ye, M.; Li, M.; Chen, Y.; Zhou, J.; Wang, L.; Liu, L. All-fiber-based photonic microwave generation with 10^{-15} frequency instability. *Chin. Opt. Lett.* **2022**, *20*, 021406. [CrossRef]
12. Fang, P.; Sun, H.; Wang, Y.; Xu, Y.; Chen, Q. Transfer of laser frequency from 729 nm to 1.5 μm with precision at the level of 10^{-20}. *Chin. Opt. Lett.* **2022**, *20*, 081403. [CrossRef]
13. Li, X.; Xu, Z.; Zhai, Y.; Yin, Y.; Ning, X.; Quan, W. Impact analysis of cavity length on transfer cavity frequency locking system for atomic inertial measurement device. *AIP Adv.* **2020**, *10*, 015340. [CrossRef]

14. Zhadnov, N.O.; Masalov, A.V. Temperature-compensated optical cavities for laser frequency stabilization. *Laser Phys. Lett.* **2023**, *20*, 030001. [CrossRef]
15. Li, G.; Duan, L.; Zhou, X.; Quan, W. Transfer matrix analysis of the birefringent fiber Fabry–Perot cavity and laser frequency locking. *Results Phys.* **2024**, *57*, 107375. [CrossRef]
16. Zhang, J.; Shi, X.H.; Zeng, X.Y.; Lu, X.L.; Deng, K.; Lu, Z.H. Characterization of electrical noise limits in ultra-stable laser systems. *Rev. Sci. Instrum.* **2016**, *87*, 123105. [CrossRef]
17. Chamoun, J.; Digonnet, M.J. Aircraft-navigation-grade laser-driven FOG with Gaussian-noise phase modulation. *Opt. Lett.* **2017**, *42*, 1600–1603. [CrossRef]
18. Ding, M.; Komanec, M.; Suslov, D.; Dousek, D.; Zvanovec, S.; Fokoua, E.R.N.; Bradley, T.D.; Poletti, F.; Richardson, D.J.; Slavik, R. Long-Length and Thermally Stable High-Finesse Fabry-Perot Interferometers Made of Hollow Core Optical Fiber. *J. Light. Technol.* **2020**, *38*, 2423–2427. [CrossRef]
19. Yang, N.; Su, J.; Fan, Z.; Qiu, Q. High Precision Temperature Insensitive Strain Sensor Based on Fiber-Optic Delay. *Sensors* **2017**, *17*, 1005. [CrossRef]
20. Hedger, J.P.; Elsmann, T.; Becker, M.; Tiess, T.; Luiten, A.N.; Sparkes, B.M. High performance fiber-Fabry-Perot resonator targeting quantum optics applications. *IEEE Photonics Technol. Lett.* **2020**, *32*, 879–882. [CrossRef]
21. Zheng, Y.X.; Cui, J.M.; Ai, M.Z.; Qian, Z.H.; Cao, H.; Huang, Y.F.; Jia, X.J.; Li, C.F.; Guo, G.C. Large-tuning-range frequency stabilization of an ultraviolet laser by an open-loop piezoelectric ceramic controlled Fabry–Pérot cavity. *Opt. Express* **2021**, *29*, 24674. [CrossRef]
22. Zou, M.; Shen, K.; Song, Q.; Dai, Y.; Xiao, X.; Sun, Q.; Yan, Z. Sub-kHz linewidth laser generation by self-injection locked distributed feedback fiber laser. *Opt. Laser Technol.* **2024**, *169*, 110022. [CrossRef]

photonics

Article

Internal Dynamic Temperature Measurement of Alkali Metal Vapor Cell by Kalman Filter

Yang Li [1], Shencheng Tian [1], Junpeng Zhao [2], Guoqing Zhou [1], Xiangmei Dong [1], Xiumin Gao [1] and Xuejing Liu [1,*]

[1] School of Optical-Electrical and Computer Engineering, University of Shanghai for Science and Technology, Shanghai 200093, China
[2] Beijing Institute of Space Launch Technology, Beijing 100076, China
* Correspondence: liuxuejing@usst.edu.cn

Abstract: Measuring the internal dynamic temperature of alkali metal vapor cells is crucial for enhancing the performance of numerous atomic devices. However, conventional methods of measuring the internal dynamic temperature of the cell are prone to errors. To obtain a more accurate internal dynamic temperature of the alkali metal vapor cell, a temperature measuring method based on the data fusion of the Kalman filter has been proposed. This method combines the indirect temperature measurement signal from a resistance temperature detector with the atomic absorption spectrometric temperature measurement signal. This provides a high-accuracy set of internal dynamic temperatures in the cell. The atomic vapor density calculated from the final fusion results is 37% average lower than that measured by external wall temperature measurements, which is in line with the conclusions reached in many previous studies. This study is highly beneficial to measure the temperature of alkali metal vapor cells.

Keywords: alkali metal vapor cell; atomic devices; atomic absorption spectrometry (AAS); Kalman filter (KF); data fusion

1. Introduction

A cell containing one or more alkali metals, inert and buffer vapors, is commonly used as a core device in atomic clocks [1], laser stabilizers [2], and optical magnetometers [3–7]. In magnetic field detection, spin-exchange-free spin (SERF) atomic magnetometers are highly valued for their ultra-high sensitivity. SERF atomic magnetometers are now widely used in biomedical [8,9] and geomagnetic exploration [10] fields. The density of alkali metal vapor in the cell is closely related to its temperature. Alcock's [11] formula for the saturation vapor pressure of alkali metals can be used to calculate the temperature. The density of alkali metal vapor significantly affects the optical depth of the cell medium, atomic polarization, atomic relaxation rate, and other factors. Therefore, exploring the internal dynamic temperature of the alkali metal cell is important.

Traditional temperature measurement of the cell involves installing a resistance temperature detector (RTD) on the outer wall of the cell, which is non-destructive [12]. However, this method has limitations. This is because the saturated vapor pressure of the cell is determined by the lowest temperature point of the entire cell, but the external temperature of the cell is not consistent, making it difficult to find the lowest temperature point of the alkali metal cell.

To address this problem many scholars have studied methods for cell temperature measurement. Shang et al. [13] used atomic absorption spectrometry (AAS) to obtain the internal temperature of the cell, which is more accurate than using RTD. Meanwhile, many scholars have found that the atomic vapor density obtained by RTD temperature using Killian's formula is much higher than the true atomic vapor density. Shao et al. [14] measured the density of alkali metal vapor, which is about half of the result calculated by the

Killian formula; Shang et al. [15] come to the same conclusion based on the measurement of alkali metal vapor density using the Faraday modulator. The reason for this phenomenon is currently unknown, but obtaining the true temperature inside the cell is an area worth exploring.

The methods for measuring the temperature of the cell are divided into two types: contact and non-contact. The non-contact measurement method is based on AAS. It offers high sensitivity, calibration-free operation, fast detection, and the ability to remotely measure the internal dynamic temperature inside the cell without entering the cell [15]. Therefore, the AAS temperature measurement method is beneficial for measuring the temperature inside the cell. RTDs have high measurement accuracy and stable performance, which is suitable for measuring temperature signals. Therefore, RTD-based temperature measurement is also suitable for measuring the temperature of cells. However, both temperature measurement methods have technical drawbacks that are hard to avoid. Measurements using RTDs require additional calculations to obtain the temperature at the center of the cell. Hence, temperature measurements using this method are inaccurate. AAS temperature measurement utilizes fitting absorption spectra, but measurement becomes challenging when the vapor density is high. This is because directly measuring the extremely weak transmitted light near resonance is difficult. Moreover, severe attenuation of light within the resonance linewidth makes it hard to determine the absorption line shape, which makes it impossible to measure through far detuned measurement [16]. Additionally, AAS measurement is affected by factors such as power fluctuations and changes in external magnetic fields. As a result, AAS may not be an accurate method for temperature measurement [17]. In this study, the results of indirect temperature measurement by RTD, temperature measurement by AAS, and direct temperature measurement of the external wall by RTD are compared. This study has found that measured temperatures vary greatly, especially during periods of increased internal dynamic temperature. The dynamic internal temperature of the cell requires a high precision measurement, but the results obtained from different temperature measurement methods may vary. Measuring the internal dynamic temperature of a cell requires high accuracy. For this purpose, a method of fusing data from two temperature measurement devices is used to measure the internal dynamic temperature of the cell. This multimodal sensor-data fusion method allows more accurate measurements [18]. Multi-sensor data fusion is useful when the measured temperature by a single sensor does not have the desired precision. So, the data from multiple sensors is processed to compensate for the individual sensor's deficiencies [19]. The data fusion methods are mainly divided into two types: modern and classical fusion methods. The modern fusion methods are based on neural networks [20], cluster analysis [21], and intelligent algorithms [22]. The classical fusion methods revolve around the Kalman filter (KF) [23] and Bayesian estimation [24].

Many research studies have been done on the KF data fusion method. Sun et al. [25] proposed an optimal information fusion dispersion KF for multiple sensors and verified the effectiveness of the method with a three-sensor radar tracking system. Wang et al. [26] fused the signals from lead-zirconium-titanium oxide (PZT) sensors and optical fiber sensors for monitoring bore edge cracks through a weighted adaptive KF. They found that the method could significantly improve the accuracy compared to single-sensor monitoring. Xu et al. [27] used KF to fuse resolution-enhanced surface temperature data with coarse-resolution surface temperature data to obtain a more accurate set of surface temperatures with a root mean square error in the range from 0.79–1.47 K. Chen et al. [28] proposed a KF-based algorithm to fuse temperature and humidity data. The experimental results showed that the algorithm filtered the data noise, reduced the data measurement error, and obtained highly accurate values.

The KF is used to fuse the measurement data from multimodal sensors. The KF acquires accurate data by combining the model state estimation with the observed data [29]. It can obtain a better set of measurements for the same object. In this study, the KF is used to measure the more accurate internal dynamic temperature of the cell. It is done by fusing correlated time series data of AAS and RTD.

The structure of this article is as follows. Firstly, the acquisition of the external wall temperature data of the vapor cell and the AAS measurement data related to the internal dynamic temperature of the cell is presented. Afterward, the data processing of the internal dynamic temperature of the cell is discussed. Next, the calculation of the internal dynamic temperature of the cell through KF data fusion for a more accurate internal dynamic temperature of the cell is done. Finally, the fusion results are compared with the external wall measurements of RTD to verify the KF fusion results. In the end, it is shown, based on the experimental results, that it is feasible to measure the internal dynamic temperature of the cell with multiple sensor data after processing with KF.

2. Measurement Procedures

2.1. AAS Temperature Measurement

Due to the lack of hyperfine pumping, hyperfine absorption is completely overlapped. Hence, the optical depth of the target can be measured according to the change in light intensity. The schematic diagram of the AAS setup is shown in Figure 1. Based on Lambert's law, the relationship between the corresponding quantities is presented in Equation (1) [30]:

$$I(O) = I(I)\exp(-OD) = I(I)\exp(-\rho \cdot l \cdot s) \tag{1}$$

where $I(I)$ is the light intensity of the laser incident on the cell, $I(O)$ is the light intensity of the laser exiting the cell, OD is the optical depth, ρ is the atomic density of the alkali metal vapor, s is the photon absorption cross-section, and l is the optical length of the laser beam in the cell.

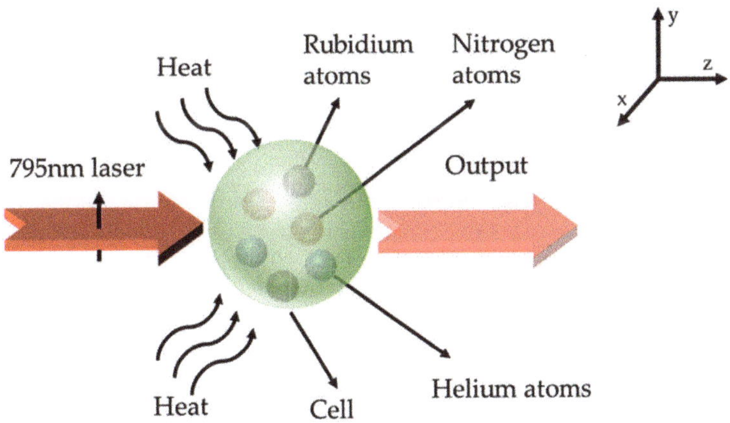

Figure 1. The schematic diagram of AAS setup. A 795 nm laser beam enters the heated cell with a light intensity *I(I)*, gets absorbed and exits the cell with a light intensity *I(O)*.

The buffer gas atom causes a linear Lorentz broadening between atoms. To avoid collision relaxation, the cell used in the experiment injects three atmospheres of ^4He. Under these conditions, an approximate broadening of 50 GHz is observed, and saturated absorption peaks were not visible. Consequently, based on a known set of measurement data, the single peak Lorentz function in Equation (2) is capable of fitting the data [31]. The fitting results are then used to calculate the full width at half maximum (FWHM) Γ, the center frequency v_0.

$$L(v - v_0) = \frac{\Gamma/(2\pi)}{(v - v_0)^2 + (\Gamma/2)^2} \tag{2}$$

Substituting Γ of the curve into Equation (3) representing the absorption area at the resonant frequency point, the photon absorption cross-sectional area $s(v)$ can be calculated as follows [32].

$$s(v) = \pi r_e c f \cdot L(v - v_0) \tag{3}$$

In Equation (3), r_e is the electron radius, c is the speed of light in vacuum, and f is the intensity of the oscillator. The atomic density n of alkali metal vapor is obtained through Equation (4), where $ODmax$ is the maximum optical depth, $s(v_0) = s_{\max}$ is the maximum photon absorption cross-sectional area at resonance [33]:

$$\rho = \frac{OD_{\max}}{s_{\max} l} \tag{4}$$

The empirical formula given in Equation (5) presents the relationship between cell temperature and the atomic density of alkali metal vapor [11,13]:

$$\rho = \frac{1}{T_G} 10^{A - \frac{B}{T_G}} \tag{5}$$

Equation (5) provides the cell temperature T_G if all the other quantities are known. A and B are the density parameters of liquid rubidium (Rb) vapor with the following values $A = 31.178$ and $B = 4040$.

2.2. RTD Indirect Temperature Measurement

The RTD temperature measurement method is invasive, which means that it needs to be in direct contact with the object being measured. However, because the cell is non-destructive and the RTD is pre-installed during cell manufacturing, the RTD may be corroded by alkali metal vapor. As a result, RTD can only measure the temperature of the outer wall of the cell.

According to Fourier's law of heat conduction, heat will move from a high temperature region to a low temperature region when there is non-uniform temperature distribution. Equation (5) indicates that the temperature of the cell is directly proportional to the atomic density of alkali metal vapor. The electric heating plate is in close contact with the cell while it is being charged and heated. The lack of internal heating in the cell results in a slight gradient of temperature along its radius. Thus, the external wall of the cell with a radius of 10 mm can be simulated as being in a state of thermal equilibrium. In the transient case of 428 K, the cell is simulated in two dimensions. In the two-dimensional case, the temperature distribution in the cell is approximated by the two-dimensional Gaussian distribution model shown in Figure 2. Therefore, in this work, a simplified temperature distribution model is proposed, in which the temperature distribution of the cell is considered to have an inverse Gaussian function.

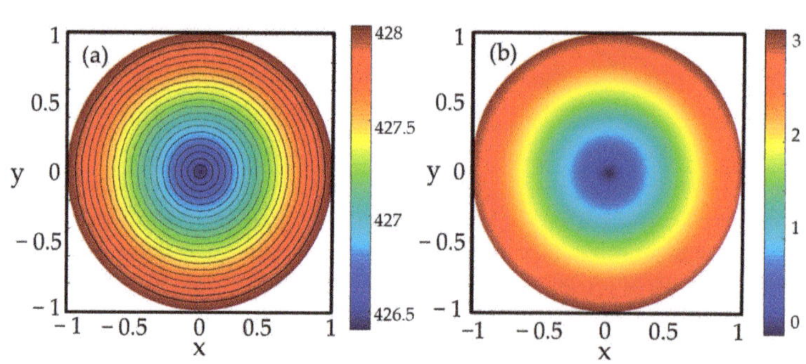

Figure 2. Cell temperature model. (**a**) The simulation plot, and (**b**) the cone distribution plot.

This article describes a simple simulation of the constant temperature situation in the 428 K cell, as shown in Figure 2a. In this case, the temperature of the cell has reached a steady state. We treat the cell wall as a constant temperature heat source, regardless of the temperature exchange between the cell and the environment.

Thus, the isotherms of the cell can be simplified into regular circles. When measured outside the cell, the extended isotherms of the heat source at any point are perpendicular to the heat transfer and, as a result, must pass through the center of the spherical cell. Additionally, assuming that heat transfer comes from an infinite number of equivalent heat sources, a way to estimate cell temperature from the temperature of the outer wall is provided.

The concentration of alkali metal vapor in the cell is determined by the point where the alkali condenses at low temperatures. However, Guo et al. [34] found that the distribution of alkali metal vapor within the cell is not uniform, and the reason for this is not fully understood. Therefore, let's consider Equation (1) and assume that the object has the same atomic density at the same temperature. In the full optical range, there are i optical lengths of varying density, and each optical length with the same density is considered a unit length. By substituting these variables in Equation (1), we can express the attenuation coefficient of laser light passing through the cell as $\lambda_1, \lambda_2, \lambda_3 \ldots\ldots \lambda_i$, and the corresponding optical length as $l_1, l_2, l_3 \cdots l_i$. The following expression is obtained:

$$\lambda_1 l_1 + \lambda_2 l_2 + \lambda_3 l_3 + \cdots + \lambda_i l_i = \ln\left(\frac{I(I)}{I(O)}\right) \tag{6}$$

The internal distribution of alkali metal vapor in the heating cell is not uneven. From Equations (4) and (5) it is clear that temperature, optical depth, and alkali metal vapor density are positively correlated. Therefore, the variation of the attenuation coefficient is also distributed in the inverted Gaussian function. Hence, Equation (7) can be used to represent the optical depth of the medium within a certain range. l_a and l_b are the lower and upper limits within the cell.

$$\int_{l_a}^{l_b} \lambda(x,y)dl = OD' \tag{7}$$

Substituting this value in Equations (4) and (5) gives the temperature T_D at a point in the interval.

2.3. KF Fusion Algorithm

The KF is a recursive optimization algorithm whose main application is in time-varying dynamic systems with uncertainty [25]. It is difficult to establish an accurate mathematical model of the internal dynamic temperature of a cell. Whether the internal dynamic temperature of the cell is extrapolated from RTD measurements or measured by AAS, there will always be unmeasurable errors. Therefore, the use of a KF for data fusion is very appropriate.

The spatial state system model of the object under test can be expressed as:

$$x(k+1) = Ax(k) + w(k) \tag{8}$$

$$y(k) = Hx(k) + v(k) \tag{9}$$

where $x(k)$ is the state matrix of the object under observation at the moment k. $y(k)$ is the state observation signal of the object under test at the moment k. $w(k)$ is the process noise at the moment k. $v(k)$ is the observation noise at the moment k. A and H are the state transfer matrix and the observation matrix, respectively. Both $w(k)$ and $v(k)$ are Gaussian white noises, and their corresponding covariance matrices are Q and R. The core formula for the KF is as follows [28].

State space projections:

$$x(k|k-1) = Ax(k-1) \tag{10}$$

$$P(k|k-1) = AP(k-1)A^T + Q \tag{11}$$

Updated state space:

$$K(k) = P(k|k-1)H^T\left[HP(k|k-1)H^T + R\right]^{-1} \tag{12}$$

$$x(k) = x(k|k-1) + K(k)[y(k) - Hx(k|k-1)] \tag{13}$$

$$P(k) = [I - K(k)H]P(k|k-1) \tag{14}$$

where $x(k\,|\,k-1)$ is the estimate obtained from the optimal estimate at $k-1$. $K(k)$ is the Kalman gain at k. $x(k)$ is the optimal estimate at k. $P(k)$ is the covariance matrix at k, and I is the unit matrix.

Data fusion with KF is the analysis of the sensor data followed by the execution of the KF algorithm to obtain an optimum estimate of the measured values. The best estimate from Equation (13) can be considered the optimal fusion of the observations and estimates consisting of time series signals collected by multiple sensors. The KF-based equation of state for the internal dynamic temperature of the cell can be expressed as Equations (15) and (16):

$$T_G(k) = \mathrm{B}_G x_G(k) + v_G(k) \tag{15}$$

$$T_D(k) = \mathrm{B}_D x_D(k) + v_D(k) \tag{16}$$

where $T_G(k)$ is the temperature measured using AAS. $T_D(k)$ is the temperature measured using RTD indirect temperature measurement method. B_G is the observation matrix for temperature measurement using AAS. B_D is the observation matrix for temperature measurement using RTD. $v_G(k)$ and $v_D(k)$ are the measurement noises for both methods. According to the core formula of the KF, a relationship given in Equation (17) exists between the two unrelated measurements.

$$T_F = \frac{D_G^{-1}T_G}{D_G^{-1} + D_D^{-1}} + \frac{D_D^{-1}T_D}{D_G^{-1} + D_D^{-1}} \tag{17}$$

where T_F is the temperature obtained after KF fusion, D_G and D_D are the variance of the two measurements, which can be obtained from multiple measurements at a fixed temperature.

To test the effectiveness of the data fusion algorithm that relied on KF, we conducted a simple simulation. During the simulation, the standard deviations of inputs A and B are 0.5 and 0.4, respectively.

Figure 3 shows the KF fusion result, demonstrating the potential temperature signals using KF fusing. The simulate output's standard deviation was reduced to 0.31, indicating the efficacy of the signal fusion algorithm using KF in temperature measurement experiments on cells.

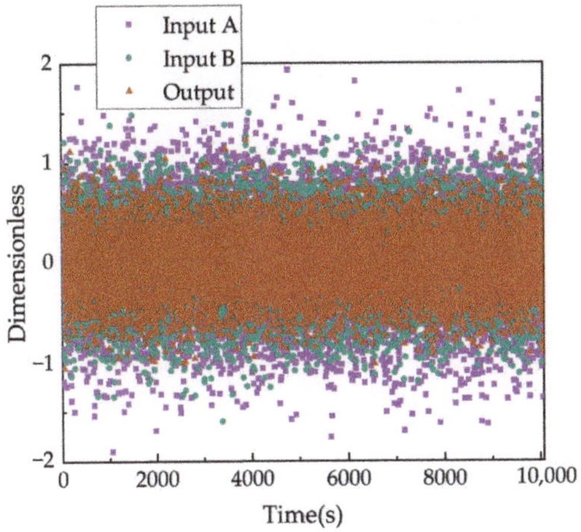

Figure 3. Simulation results of a signal fusion using a KF. It is evident that the variance of the output signal is significantly lower than that of the two input signals.

3. Experimental Setups

The experimental setup consists of a Rb atomic cell, heating oven, electric heating plate, a computer, UniQuanta laser (DFB801-795), two photodetectors (S142C), optical power meter (PM100USB), RTD data acquisition card (M2111) and four RTDs. The schematic diagram of the experimental setup is shown in Figure 4. A 20 mm diameter spherical borosilicate glass cell containing Rb atoms, quenching gas N_2, and buffer gas ^4He is placed in an oven made of boron nitride. The heating capacity of the oven is provided by two non-magnetic, temperature-controlled electric heating plates with a maximum heating capacity of 473 K. N_2, and ^4He gases do not produce absorption peaks in the visible light band. In the visible light band, the wavelength of Rb atomic line is at about 794.98 nm. Therefore, the N_2 and ^4He gases do not affect the experiment. Four RTDs are attached to the outer wall of the cell on both sides of the two ventilation holes. The RTDs are small enough (2 × 5 mm) not to obstruct the transmission of light. This ensures that the RTDs can obtain the most direct temperature of the outer wall of the cell. This also prevents experimental failure due to the malfunction of the small RTDs during experiments. The temperature value measured by the RTD is the average of four individual temperature measurements.

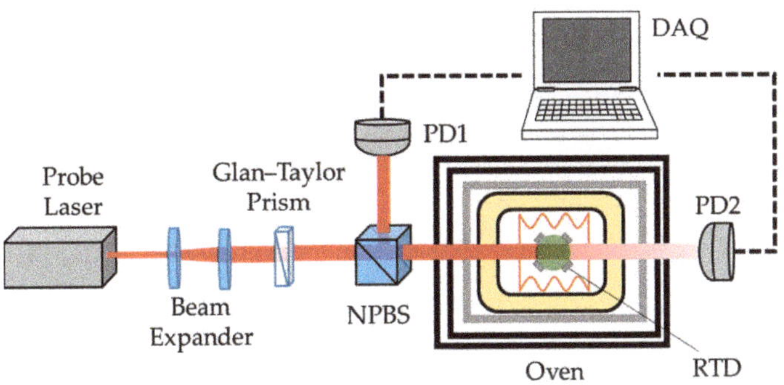

Figure 4. Schematic diagram of the internal dynamic temperature measurement test platform inside the cell.

The heated oven is packed in a non-magnetic tank with two ventilation holes. The detection light enters through a vent on one side of the tank, passes through the cell, and exits through a vent on the other side. A distributed feedback laser supplies the detection light with a central wavelength of 795 nm. After passing through two plano-convex lenses, the beam is transformed into a linearly polarized beam by using a Glan-Taylor prism as a polarizer. The linearly polarized beam is split in two with a 1:1 ratio using a non-polarization beam splitter (NPBS). One of the beams is directly captured by the photodetector PD1 to detect the intensity of the light. The other beam is captured by the photodetector PD2 after passing through the oven and being absorbed by the Rb atoms to detect the intensity of the light.

Since the cell needs to reach a specific temperature before it can start working, the data for this experiment is recorded from when the temperature of the cell reached 373 K and ended when the external temperature control section of the cell showed a stable temperature of 428 K. A total of 10,800 sampling points were recorded.

Finally, two sets of dynamic internal cell temperatures were obtained for data fusion. Another data set, the external wall's internal dynamic temperature, measured directly, was used for subsequent comparison.

4. Results and Discussion

To measure temperature using AAS, it is necessary to scan DFB lasers near the Rb atomic D1 line to obtain optical depths at different laser frequencies. This is required due to the optical thickness that occurs at higher temperatures and drifts in the DFB laser during usage, which makes it difficult to measure temperature at a fixed frequency. To conduct the experiment, the oven must first be set to a certain temperature. After the oven has reached the desired temperature and has stabilized within a fluctuation range of ±0.1 K, adjust the wavelength of the DFB laser to scan near the Rb atom's D1 line. It is important to note that the wavelength adjustment step size of the DFB laser varies. Near the D1 line, the step size is 0.08 nm, while it is 0.16 nm, 0.24 nm, and 0.4 nm away from the D1 line. This process was repeated at each working temperature of the cell to calibrate the temperature measurement by AAS.

The OD obtained by AAS varies at different temperatures. This is because, as the cell temperature increases, $I(O)$ gradually decreases while $I(I)$ remains constant. To ensure consistent evaluation of the residual sum of squares (RSS) evaluation indicators for different sets of data after fitting, OD was normalized. RSS is a metric that measures the sum of squares between the actual value OD_i^* and the fitted value $\overset{\frown}{OD}_i$. Usually lower values indicates a better fit. The expression for RSS can be found in Equation (18).

$$RSS = \sum\left(OD_i^* - \overset{\frown}{OD}_i\right)^2 \tag{18}$$

Normalization does not affect the data wavelength and therefore does not affect the FWHM obtained after fitting. In this experiment, extreme difference normalization was selected, as shown in Equation (19):

$$OD_i^* = \frac{OD_i - \min OD_i}{\max OD_i - \min OD_i} \tag{19}$$

Temperature measurements by AAS were carried out in 5 K intervals in the operating temperature range of 333–428 K inside the Rb cell. The optical depth as a function of wavelength is obtained over the wavelength range of 794–796 nm in various intervals, as shown in Figure 5, which illustrates the absorption peaks at each sampling temperature.

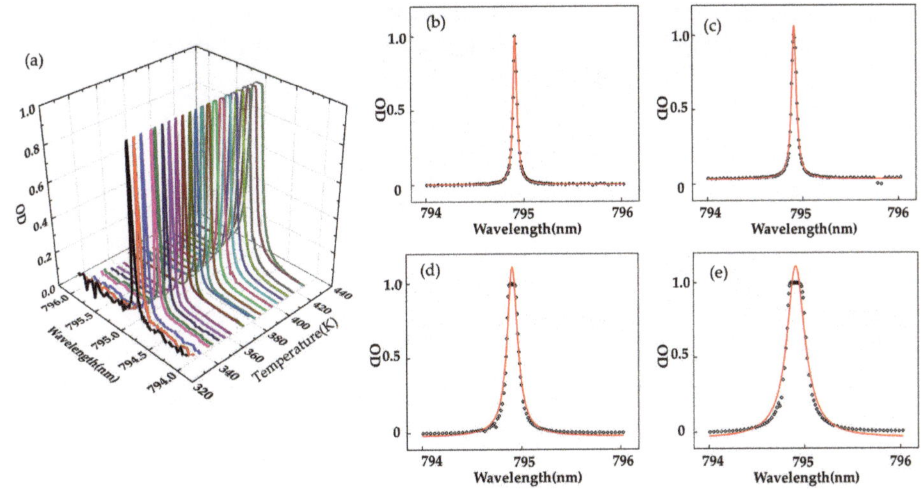

Figure 5. Raw data and fitting results of AAS measurements at each temperature. In items (**b–e**), the dots represent the measurement results of laser at different wavelengths at a fixed temperature, while the red line represents the fitting curve at the fixed temperature. (**a**) Absorption peaks at each sampling temperature in the wavelength range from 794–796 nm. (**b**) Lorentzian fit at 358 K. (**c**) Lorentzian fit at 378 K. (**d**) Lorentzian fit at 398 K. (**e**) Lorentzian fit at 418 K.

The results in Figure 5a indicate that the absorption peak exhibits significant fluctuations on both sides below 333 K, which could be attributed to the incomplete work of the cell. Moreover, Figure 5b–e demonstrates that the absorption peaks become more optically thick as the temperature increases. The waveforms of the absorption peaks also gradually become irregular from the Lorentzian type, resulting in incomplete fitting. An optically thick phenomenon is evident after 398 K.

Table 1 shows that after 11 iterations of fitting with the Lorentz function, the raw data obtained with AAS provides a better fit at low temperatures, but a poorer fit at high temperatures. Inaccuracies in the FWHM obtained by the Lorentz fit can occur due to lower fits, resulting in deviations in calculated temperatures. The high atomic density of the alkali metal vapor at higher temperatures results in complete absorption of the incident beam, ultimately leading to an optically thick absorption peak. In addition, when the laser emitted from the cell is very small, the limited sensitivity of the photodetector resulting in intensity values is unmeasurable, leading to flat peaks. This variance is significant because the indirect RTD measurement method calculates the internal dynamic temperature of the cell using the external temperature of the cell measured by the RTD, and Equation (7) is used in this process. It can seriously affect the accurate measurement of the internal dynamic temperature of the cell.

Table 1. Lorentz function fitting at 358 K, 378 K, 398 K, and 418 K.

Temperature	358 K		378 K		398 K		418 K	
	RSS	R^2	RSS	R^2	RSS	R^2	RSS	R^2
	0.02576	0.9932	0.01307	0.99729	0.14769	0.98528	0.32702	0.97621

To ensure the accuracy of the data fusion via KF, the oven was cooled to room temperature and reheated to 373 K. Afterwards, both temperature measurement methods were repeated once the temperature stabilized, with fluctuations around ±0.1 K. The purpose of these measurements was to calculate the variance of the respective methods. The final variance values are $D_G = 0.0517$ and $D_D = 0.1059$. Subsequently, the data from two temperature measurement methods were fused by KF. The results are displayed in Figure 6.

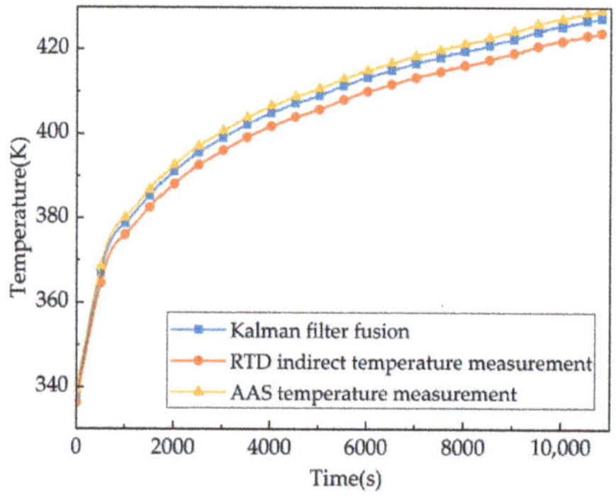

Figure 6. Temperature measurement results using Kalman filter fusion, RTD indirect temperature measurement and AAS temperature measurement.

Figure 6 indicates that the KF fused data is not as high or low as the input data. This is because the output data combines the characteristics of the two input data sets using KF. In this study, the non-magnetic electric heating plates method was used to increase the temperature. However, this method resulted in a sudden temperature rise, most likely due to frequency variation of the plates. This sudden increase in temperature after KF data fusion cannot be eliminated. The consistency of the three curves is high because the measurement data of AAS used RTD measurement data as a reference, resulting in the same trend of the two sets of data. The KF fusion results calculated by Equation (17) also obtained this trend information, thereby ensuring higher consistency.

As shown in Figure 7, the atomic vapor density calculated from the temperature values obtained by fusing the KF data is approximately 37% lower on average than the corresponding values from conventional external wall temperature measurements. This finding is consistent with the results of other studies on calculating atomic vapor density [14,15,35,36]. Furthermore, the density of atomic vapor within the cell becomes progressively more homogeneous as the temperature increases. This is due to the gradual increase of the ratio between the two results. However, after 10,000 s, the temperature increase reaches a state of stagnation. Even when the temperature is stabilized, there is still a 20% difference between the measurement results of the two methods, which may be attributed to the inhomogeneous temperature of the cell.

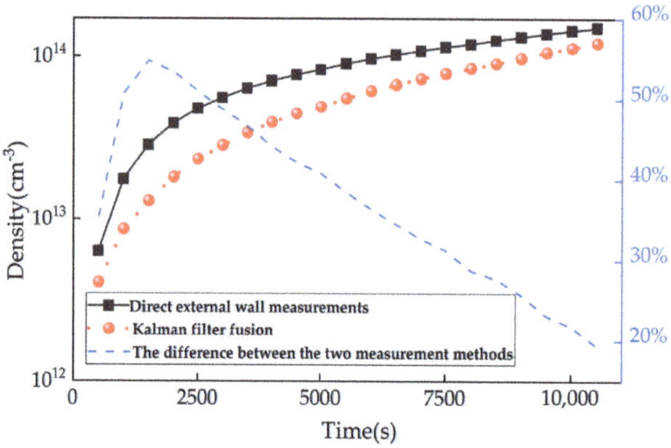

Figure 7. The values of atomic vapor density by KF data fusion and conventional direct gas method.

5. Conclusions

In this study, we used the KF data fusion algorithm to fuse data from RTD and AAS to measure cell temperature more precisely, which is typically difficult. We applied the algorithm to experimental data and observed an improvement in the output result. The experimental results indicate that the proposed KF fusion method can accurately measure the internal dynamic temperature of the cell. In addition, the atomic vapor density calculated by the KF fusion method is approximately 37% lower than the corresponding values directly measured from the outer wall of the cell. This study can be beneficial in improving the accuracy of cell temperature measurement during heating and enhancing the performance of the related atomic precision instrumentation.

However, the low sensitivity of the photodetector used in this study limited the precision of the measurements in the AAS method, specifically for high-temperature measurements. In future work, we could consider using the Faraday modulator [15,37] to address the issue of optical thickness and improve the optical path to enhance the performance of the equipment for the improvement of this study.

Author Contributions: Conceptualization, S.T.; methodology, Y.L.; software, Y.L.; validation, Y.L.; formal analysis, S.T.; investigation, X.L.; resources, Y.L.; data curation, S.T.; writing—original draft preparation, Y.L.; writing—review and editing, J.Z.; visualization, G.Z.; supervision, X.L.; project administration, X.D.; funding acquisition, Y.L and X.G. All authors have read and agreed to the published version of the manuscript.

Funding: This research was funded by the National Natural Science Foundation of China, grant number 52202456.

Institutional Review Board Statement: Not applicable.

Informed Consent Statement: Not applicable.

Data Availability Statement: Not applicable.

Conflicts of Interest: The authors declare no conflict of interest.

References

1. Zhang, S.W.; Lu, J.X.; Ye, M.; Zhou, Y.; Yin, K.F.; Lu, F.; Yan, Y.G.; Zhai, Y.Y. Optimal Operating Temperature of Miniaturized Optically Pumped Magnetometers. *IEEE Trans. Instrum. Meas.* **2022**, *71*, 7. [CrossRef]
2. Affolderbach, C.; Mileti, G. A compact laser head with high-frequency stability for Rb atomic clocks and optical instrumentation. *Rev. Sci. Instrum.* **2005**, *76*, 073108. [CrossRef]
3. Boto, E.; Meyer, S.S.; Shah, V.; Alem, O.; Knappe, S.; Kruger, P.; Fromhold, T.M.; Lim, M.; Glover, P.M.; Morris, P.G.; et al. A new generation of magnetoencephalography: Room temperature measurements using optically-pumped magnetometers. *NeuroImage* **2017**, *149*, 404–414. [CrossRef] [PubMed]
4. Miao, P.X.; Zheng, W.Q.; Yang, S.Y.; Wu, B.; Cheng, B.; Tu, J.H.; Ke, H.L.; Yang, W.; Wang, J.; Cui, J.Z.; et al. Wide-range and self-locking atomic magnetometer based on free spin precession. *J. Opt. Soc. Am. B* **2019**, *36*, 819–828. [CrossRef]
5. Lu, J.X.; Qian, Z.; Fang, J.C.; Quan, W. Suppression of vapor cell temperature error for spin-exchange-relaxation-free magnetometer. *Rev. Sci. Instrum.* **2015**, *86*, 083103. [CrossRef]
6. Auzinsh, M.; Sargsyan, A.; Tonoyan, A.; Leroy, C.; Momier, R.; Sarkisyan, D.; Papoyan, A. Wide range linear magnetometer based on a sub-microsized K vapor cell. *Appl. Opt.* **2022**, *61*, 5749–5754. [CrossRef]
7. Yan, Y.G.; Lu, J.X.; Zhou, B.Q.; Wang, K.; Liu, Z.A.; Li, X.Y.; Wang, W.Y.; Liu, G. Analysis and Correction of the Crosstalk Effect in a Three-Axis SERF Atomic Magnetometer. *Photonics* **2022**, *9*, 654. [CrossRef]
8. Ma, Y.T.; Qiao, Z.X.; Yu, M.Z.; Wang, Y.B.; Chen, Y.; Luo, G.X.; Yang, P.; Lin, Q.J.; Zhao, L.B.; Zhang, Y.; et al. Single-beam integrated hybrid optical pumping spin exchange relaxation free magnetometer for biomedical applications. *Appl. Phys. Lett.* **2022**, *121*, 114001. [CrossRef]
9. Ma, Y.T.; Chen, Y.; Zhao, L.B.; Yu, M.Z.; Wang, Y.B.; Guo, J.; Yang, P.; Lin, Q.J.; Jiang, Z.D. Accurate determination of alkali atom density based on zero-field magnetic resonance in a single-beam spin-exchange relaxation-free atomic magnetometer. *Meas. Sci. Technol.* **2022**, *33*, 105003. [CrossRef]
10. Ma, Y.N.; Lu, J.X.; Yang, K.; Wang, Z.X.; Wang, Y.G.; Sun, B.W.; Zhai, Y.Y.; Han, B.C. Measuring Spin Polarization of a Spin-Exchange Relaxation-Free Atomic Magnetometer at Extremely Large Optical Depths. *IEEE Trans. Instrum. Meas.* **2022**, *71*, 7007209. [CrossRef]
11. Alcock, C.B.; Itkin, V.P.; Horrigan, M.K. Vapour pressure equations for the metallic elements: 298–2500 K. *Can. Metall. Q.* **1984**, *23*, 309–313. [CrossRef]

12. Lu, J.X.; Wang, J.; Yang, K.; Zhao, J.P.; Quan, W.; Han, B.C.; Ding, M. In-Situ Measurement of Electrical-Heating-Induced Magnetic Field for an Atomic Magnetometer. *Sensors* **2020**, *20*, 1826. [CrossRef]
13. Shang, H.; Zou, S.; Quan, W.; Zhou, B.; Li, S.; Zhou, W.; Zhao, F. Design of a Measuring Device and Experimental Study into the Relationship between Temperature and the Density of Alkali Metal-Vapor. *Photonics* **2023**, *10*, 112. [CrossRef]
14. Shao, W.; Wang, G.; Hughes, E.W. Measurement of spin-exchange rate constants between ^{129}Xe and alkali metals. *Phys. Rev. A* **2005**, *72*, 022713. [CrossRef]
15. Shang, H.; Zhou, B.; Quan, W.; Chi, H.; Fang, J.; Zou, S. Measurement of rubidium vapor number density based on Faraday modulator. *J. Phys. D* **2022**, *55*, 335106. [CrossRef]
16. Wei, K.; Zhao, T.; Fang, X.; Zhai, Y.; Li, H.; Quan, W. In-situ measurement of the density ratio of K-Rb hybrid vapor cell using spin-exchange collision mixing of the K and Rb light shifts. *Opt. Express* **2019**, *27*, 16169–16183. [CrossRef]
17. Yin, Y.; Zhou, B.Q.; Wang, Y.X.; Ye, M.; Ning, X.L.; Han, B.C.; Fang, J.C. The influence of modulated magnetic field on light absorption in SERF atomic magnetometer. *Rev. Sci. Instrum.* **2022**, *93*, 6. [CrossRef]
18. Gravina, R.; Alinia, P.; Ghasemzadeh, H.; Fortino, G. Multi-sensor fusion in body sensor networks: State-of-the-art and research challenges. *Inf. Fusion* **2017**, *35*, 68–80. [CrossRef]
19. Xiao, F. Evidence combination based on prospect theory for multi-sensor data fusion. *ISA Trans.* **2020**, *106*, 253–261. [CrossRef]
20. Li, W.; Cui, Z.; Jin, M.; Yu, S.; Wang, X. Dynamic temperature measurement with a dual-thermocouple sensor based on a dual-head one-dimensional convolutional neural network. *Measurement* **2021**, *182*, 109679. [CrossRef]
21. Apiletti, D.; Baralis, E.; Bruno, G.; Cerquitelli, T. Real-time analysis of physiological data to support medical applications. *IEEE. Trans. Biomed. Eng.* **2009**, *13*, 313–321. [CrossRef] [PubMed]
22. Dameshghi, A.; Refan, M.H. Wind turbine gearbox condition monitoring and fault diagnosis based on multi-sensor information fusion of SCADA and DSER-PSO-WRVM method. *Int. J. Model. Simul.* **2019**, *39*, 48–72. [CrossRef]
23. Gan, Q.; Harris, C.J. Comparison of two measurement fusion methods for Kalman-filter-based multisensor data fusion. *IEEE. Trans. Aerosp. Electron. Syst.* **2001**, *37*, 273–279. [CrossRef]
24. Slimani, A.; Ribot, P.; Chanthery, E.; Rachedi, N. Fusion of model-based and data-based fault diagnosis approaches. *IFAC-PapersOnLine* **2018**, *51*, 1205–1211. [CrossRef]
25. Sun, S.L.; Deng, Z.L. Multi-sensor optimal information fusion Kalman filter. *Automatica* **2004**, *40*, 1017–1023. [CrossRef]
26. Wang, Y.; He, M.; Sun, L.; Wu, D.; Wang, Y.; Qing, X. Weighted adaptive Kalman filtering-based diverse information fusion for hole edge crack monitoring. *Mech. Syst. Signal. Process* **2022**, *167*, 108534. [CrossRef]
27. Xu, S.; Cheng, J. A new land surface temperature fusion strategy based on cumulative distribution function matching and multiresolution Kalman filtering. *Remote Sens. Environ.* **2021**, *254*, 112256. [CrossRef]
28. Chen, Y.; Xu, J.; Luo, K.; Xu, S. Multi-temperature and humidity data fusion algorithm based on Kalman filter. *Res. J. Appl. Sci. Eng. Technol.* **2013**, *5*, 2127–2132. [CrossRef]
29. Kalman, R. A new approach to linear filtering and prediction problems. *J. Basic Eng.* **1960**, *82*, 35–45. [CrossRef]
30. Mäntele, W.; Deniz, E. UV–VIS absorption spectroscopy: Lambert-Beer reloaded. *Spectrochim. Acta A Mol. Biomol. Spectrosc.* **2017**, *173*, 965–968. [CrossRef]
31. Seltzer, S.J. Developments in Alkali-Metal Atomic Magnetometry. Ph.D. Thesis, Department of Physics, Princeton University, Princeton, NJ, USA, 2008; pp. 238–240.
32. Kluttz, K.A.; Averett, T.D.; Wolin, B.A. Pressure broadening and frequency shift of the D1 and D2 lines of Rb and K in the presence of He3 and N2. *Phys. Rev. A* **2013**, *87*, 032516. [CrossRef]
33. Zheng, H.J.; Quan, W.; Liu, X.; Chen, Y.; Lu, J.X. Measurement of Atomic Number of Alkali Vapor and Pressure of Buffer Gas Based on Atomic Absorption. *Spectrosc. Spectr. Anal.* **2015**, *35*, 507–511.
34. Guo, Q.; Li, X.; Zhang, N.; Lu, J.; Ma, D.; Li, Z.; Jiang, S. The Space Density Distribution of Alkali Metal Atoms in a SERF Atomic Magnetometer. *IEEE Sens. J.* **2022**, *22*, 6476–6481. [CrossRef]
35. Chann, B.; Babcock, E.; Anderson, L.W.; Walker, T.G. Measurements of He-3 spin-exchange rates. *Phys. Rev. A* **2002**, *66*, 032703. [CrossRef]
36. Kadlecek, S.; Anderson, L.W.; Walker, T. Measurement of potassium-potassium spin relaxation cross sections. *Nucl. Instrum. Methods Phys. Res.* **1998**, *402*, 208–211. [CrossRef]
37. Shang, H.N.; Quan, W.; Chen, Y.; Li, Y.; Li, H. The Measuring Method of Atomic Polarization of Alkali Metal Vapor Based on Optical Rotation and the Analysis of the Influence Factors. *Spectrosc. Spectr. Anal.* **2016**, *36*, 305–309.

photonics

Article

The Optimization of Microwave Field Characteristics for ODMR Measurement of Nitrogen-Vacancy Centers in Diamond

Zhenxian Fan [1,2], Li Xing [2,*], Feixiang Wu [3], Xiaojuan Feng [2,*] and Jintao Zhang [2]

1 Department of Precision Instrument, Tsinghua University, Beijing 100084, China; fanzx20@tsinghua.org.cn
2 National Institute of Metrology, Beijing 100029, China; zhangjint@nim.ac.cn
3 College of Mechanical and Electrical Engineering, China Jiliang University, Hangzhou 310018, China; s22010811037@cjlu.edu.cn
* Correspondence: xingli@nim.ac.cn (L.X.); fengxj@nim.ac.cn (X.F.)

Abstract: A typical solid-state quantum sensor can be developed based on negatively charged nitrogen-vacancy (NV^-) centers in diamond. The electron spin state of NV^- can be controlled and read at room temperature. Through optical detection magnetic resonance (ODMR) technology, temperature measurement can be achieved at the nanoscale. The key to ODMR technology is to apply microwave resonance to manipulate the electron spin state of the NV^-. Therefore, the microwave field characteristics formed near the NV^- have a crucial impact on the sensitivity of ODMR measurement. This article mainly focuses on the temperature situation in cellular applications and simulates the influence of structural parameters of double open loop resonant (DOLR) microwave antennas and broadband large-area (BLA) microwave antennas on the microwave field's resonance frequency, quality factor Q, magnetic field strength, uniformity, etc. The parameters are optimized to have sufficient bandwidth, high signal-to-noise ratio, low power loss, and high magnetic field strength in the temperature range of 36 °C to 42.5 °C. Finally, the ODMR spectra are used for effect comparison, and the signal-to-noise ratio and Q values of the ODMR spectra are compared when using different antennas. We have provided an optimization method for the design of microwave antennas and it is concluded that the DOLR microwave antenna is more suitable for living cell temperature measurement in the future.

Keywords: NV^- centers in diamond; microwave field characteristics; microwave antenna; living cell temperature measurement

1. Introduction

Temperature is a fundamental factor reflecting the physical and chemical changes in living organisms. For example, living cells can cause internal temperature changes when responding to environmental changes [1,2], but the temperature changes generated by this process are usually small and brief [3], posing a challenge to temperature measurement technology. In recent years, existing intracellular temperature measurement technologies, such as fluorescent proteins [4], organic dyes [5], and rare earth particles [6], have generally suffered from low measurement accuracy [7], insufficient temporal and spatial resolution, and unstable fluorescence [8]. The method of NV centers in diamond stands out due to its stable physical and chemical properties, good biocompatibility [9,10], and ultra-high sensitivity in electronic spin measurement [11]. Meanwhile, it can measure multiple physical quantities such as temperature [12–14], magnetic field [15,16], pressure [17,18], electric field [19,20], etc. Therefore, it has become an emerging micro–nanoscale quantum measurement technology, especially in the fields of temperature and magnetic measurement. In 2010, the principle of temperature measurement was first proposed by D. Budker et al. [21] and the sensitivity of zero field splitting energy D to temperature was discovered. In 2021, the sensitivity (the noise spectral density) reached the level of 76 $\mu K/\sqrt{Hz}$ [7]. The NV center sensing technology requires microwave tuning, and the microwave field characteristics will directly affect the measurement signal-to-noise ratio. Microwave fields can be generated through different types of microwave antennas, among which the microwave field generated by straight copper wire antennas has poor uniformity, and the field strength decreases rapidly with increasing distance from the wire. The microwave field strength generated by a circular antenna is more uniform, but the field strength is often much weaker than that of a single copper wire [22]. For a resonator antenna, the shape, size, and thickness of the surface copper will determine the resonance frequency and magnetic field uniformity.

Therefore, in recent years, domestic and foreign research groups have designed various microwave antennas to meet the different requirements of NV center thermometers applied in different scenarios. In 2014, K. Bayat et al. designed a double open loop resonant (DOLR) microwave antenna with a resonance frequency near 2.87 GHz at room temperature. The antenna has a good quality factor, narrow bandwidth, and is suitable for ODMR spectral line measurement with a single resonant peak [23]. In 2015, D. S. Rudnicki et al. designed a microwave antenna that can adjust the resonance frequency between 2.7 GHz and 3.1 GHz, corresponding to a magnetic field of 0-100 Gauss, allowing the microwave to be arbitrarily adjusted between linear polarization and circular polarization. However, the distance between microstrip lines must match the distance between the microstrip line and the excitation point in order to obtain a good polarization state, resulting in strong position dependence and high calibration difficulty [24]. In 2016, K. Sasaki et al. designed a microwave antenna that operates near 2.87 GHz at room temperature, with a bandwidth of around 400 MHz. It can be directly applied to observe multiple Zeeman level splitting peaks under an external magnetic field, without readjusting the antenna's resonance frequency [22]. Qin et al. compared the performance between copper wire and ring microstrip antennas in 2018 [25]. After 2018, research on antennas mainly focused on optimizing uniformity in three-dimensional space. Among them, Tang Jun's team [26] designed a large-area three-dimensional uniform microwave antenna, which can increase the bandwidth and improve the contrast of ODMR. Multiple resonance frequencies can be manipulated and the antenna can be kept away from the sample without interfering with the objective lens or sample by increasing the signal-to-noise ratio (SNR) of the antenna. Moreover, the impact of microwave heating on diamond can be suppressed and quality factors can be improved. In the same year, Chu-Feng Liu's team [7] designed a dielectric resonator antenna for three-dimensional uniform operation of NV centers, achieving a Rabi frequency of 10 MHz in a 7 mm^3 space, which can increase the sensitivity of magnetometers by two orders of magnitude. Fujiwara and Shikano also reviewed the different kinds of antennas [27].

At present, the design of microwave antennas is mainly aimed at applications at room temperature or situations with external magnetic fields. However, there is a lack of further analysis on antennas used in cellular environments. On the basis of existing antenna design, this study conducts parameter optimization design research to meet the requirements of microwave fields in cell experiments. In recent years, the electromagnetic properties of cells and molecules have also received attention with the deepening of interdisciplinary research. The configuration of culture dishes [28], dielectric parameters, and electromagnetic parameters of culture media can all affect the microwave field characteristics, thereby affecting the temperature measurement of NV centers. The temperature range of the cell activity research is 36~42.5 °C, that is, from the optimal culture temperature for live cells to the temperature at which tumor cell are killed. The corresponding frequencies of zero field splitting energy are about 2.86892~2.86844 GHz. The microwave antenna itself has a more significant thermal effect compared with the relatively small changes in the thermal environment of cells. Thus, the influence of microwave thermal effect needs to be considered, and it was first observed in the study of Fujiwara et al. [29]. However, it only qualitatively revealed the existence of this phenomenon and its impact on temperature measurement accuracy. Quantitative research can be found in Ref. [30], which showed that the temperature rise caused by microwave thermal effect increases with the increase in the application time and power. At a microwave power of 6 dBm, the maximum temperature rise can reach 43.68 K. The temperature difference that living cells can adapt to is only about 5.5 K. The heat generated by microwaves is transferred to cells through the culture medium, which not only affects the accuracy of temperature measurement but also causes cell death. In summary, the requirements for microwave field characteristics in living cell temperature measurement mainly include: 1. High Q value and low heat production; 2. low return loss and high energy transmission efficiency of microwave antennas; 3. bandwidth and magnetic field uniformity meet requirements but are not demanding.

This article describes in-depth research on the microwave characteristics of two types of irregular microwave antennas, double open loop resonance (DOLR) and broadband and large-area (BLA) microwave antennas, in the temperature measurement of living cells. Firstly, the characteristics of microwave fields required for cell temperature measurement have been analyzed. Secondly, the characteristics of antenna microwave reflection parameters (S11), magnetic field strength and uniformity, as well as quality factor have been simulated, and parameter optimization methods were studied. Finally, ODMR spectral line tests were conducted to compare performance of a straight copper wire (SCW) antenna and the optimized irregular antenna. The improvement of microwave field characteristics was verified through changes in spectral line contrast, line width, and signal-to-noise ratio.

2. The Fundamental Principles

Diamond crystal is a tetrahedron composed of carbon atoms, each of which forms a covalent bond with the other four carbon atoms in a sp^3 hybrid orbital, exhibiting high symmetry [31]. When one of the carbon atoms is replaced by a nitrogen atom and the adjacent carbon atom is replaced by a vacancy, a nitrogen vacancy center is formed. According to whether they are negatively charged or not, the nitrogen vacancy centers can be divided into two types: NV^0 and NV^-, which can be converted to each other under certain environmental conditions [32,33]. The outermost layer of NV^- has two free electrons that can generate zero field splitting through spin–spin interaction. The zero field splitting energy D is temperature dependent, so temperature measurement can be achieved through spin manipulation.

As shown in Figure 1, the ground state of the NV center includes $m_s = \pm 1$ and $m_s = 0$. The zero field splitting D of the ground state electronic spins at room temperature is about 2.87 GHz. ODMR spectral lines can be obtained by scanning the microwave frequency and collecting the fluorescence [34–36]. When the NV color center is excited from the ground state to the excited state, there are two ways of de-excitation. One is spontaneous emission transition, which mainly occurs in the electronic spins in the excited state of $|0\rangle$. The electronic spin in this process directly returns from the excited state to the ground state and radiates photons. Another type is intersystem crossing (ISC), which mainly occurs when the electronic spin is in the excited state of $|\pm 1\rangle$. This process reaches a metastable state before returning to the ground state without radiating photons. From the perspective of fluorescence brightness, $|0\rangle$ is in the bright state and $|\pm 1\rangle$ is in the dark state. The function of microwaves is to manipulate the electronic spins between the $|0\rangle$ state and the $|\pm 1\rangle$ state, causing population flipping. When the microwave frequency is the same as the frequency of zero field splitting energy D, the population of electronic spins in the $|\pm 1\rangle$ state is the highest, resulting in the lowest fluorescence emitted by the NV^-. Moreover, the variation of microwave power can cause a change in the population of electron spin in the $|\pm 1\rangle$ state, manifested as a variation in ODMR contrast. In addition, changes in microwave power can cause changes in the spectral line width [27]. The contrast C and line width $\Delta\upsilon$ directly affect the theoretical signal-to-noise ratio of temperature measurement $\eta_T(\mathrm{K}/\sqrt{\mathrm{Hz}})$, i.e., $\eta_T(\mathrm{K}/\sqrt{\mathrm{Hz}}) \approx P_F(\frac{dT}{dv})_{T_0}\Delta\upsilon/(C\sqrt{R})$. It is known that the larger the contrast C and the narrower the line width $\Delta\upsilon$, the lower the noise spectral density. Therefore, the design of microwave antennas is crucial.

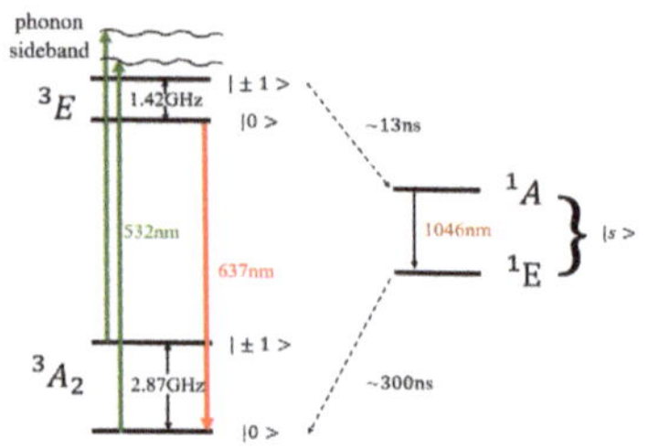

Figure 1. The energy level structure of NV^- center without considering external magnetic field and stress.

The DOLR and BLA microwave antennas studied by K. Bayat [23] and K. Sasaki [22] are suitable for room temperature applications, as shown in Figure 2. The impact of culture dishes and media on the performance of microwave antennas in cell temperature measurement was not considered. Therefore, it is necessary to optimize the structural parameters to make them suitable for cell research scenarios.

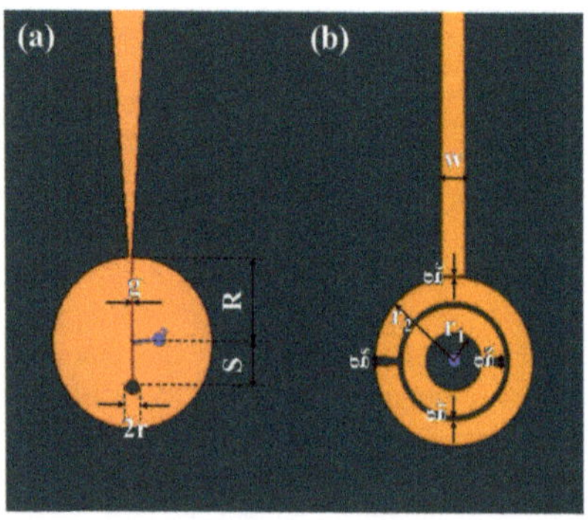

Figure 2. Schematic diagram of two types of irregular microwave antennas. (**a**) BLA microwave antenna structure; (**b**) DOLR microwave antenna structure.

3. Simulation and Optimization of Irregular Antennas

3.1. Microwave Radiation Element Simulation

Firstly, numerical simulation was conducted based on the microwave antenna structural parameters set in the K. Bayat [23] and K. Sasaki [22] papers. The simulation results are analyzed from the aspects of reflection parameter *S*11, quality factor *Q*, magnetic field strength, uniformity, etc.

For the center frequency of microwave antennas, that of the BLA microwave antenna is 2.9680 GHz, which is 0.098 GHz larger than 2.87 GHz with a frequency deviation of about 3.4%. The center frequency of the DOLR microwave antenna is 2.8797 GHz, which is close to the microwave frequency corresponding to zero field splitting energy at room temperature, with a frequency deviation of only 0.3%. The center frequency of the two types of antennas can be adjusted later by pasting copper strips. The bandwidth of BLA microwave antennas can reach 400 MHz, ensuring that Zeeman level splitting lines can be observed under an external magnetic field of up to 100 Gauss. For DOLR microwave antennas, the bandwidth is only a few tens of megahertz, making it difficult to measure multiple resonance peaks simultaneously within the bandwidth range. However, for the temperature measurement of the cell environment, the D value changes by about 400 kHz, and the bandwidth requirement is not strict. The reflection parameter S11 of the two antennas is shown in Figure 3. The S11 of the DOLR is about −9 dBm and that of BLA is about −5 dBm, thus the return loss of the DOLR microwave antenna is relatively small and the power transmission efficiency will be higher. The quality factor Q is the ratio of the electromagnetic energy stored in the resonator to the energy lost in one cycle, which is related to the ohmic loss of the metal wire, the dielectric loss of the printed circuit board, the radiation loss, and the coupling loss of the microstrip line. Its value can be obtained by the ratio of center frequency to line width. The microwave field characteristics of the two antennas are listed in Table 1, and the Q value of DOLR is 18 times larger than that of BLA. In summary, the DOLR microwave antenna is more suitable for situations where the antenna heating effects need to be weakened and measured at the nanoscale.

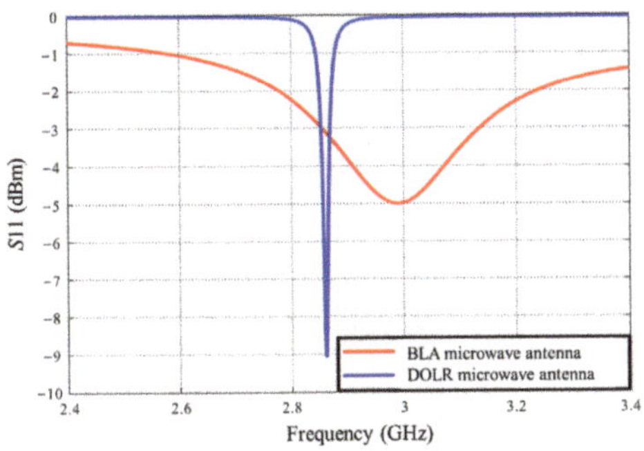

Figure 3. Reflection parameter S11 spectra of two types of microwave antennas.

Table 1. Microwave Field Characteristics of Microwave Antennas.

Microwave Field Characteristics \ Antenna Type	BLA Microwave Antenna		DOLR Microwave Antenna
Center frequency (GHz)	2.968		2.880
Bandwidth (MHz)	379		19
Q value	7.8		151.6
Uniformity of microwave field on different areas of microwave antennas ($H'_{homogeneity} = (H_{max} - H_{min})/H_{mid}$)	4 mm × 4 mm	1.95	1.80
	3 mm × 3 mm	1.89	1.75
	2 mm × 2 mm	1.80	1.27
	1 mm × 1 mm	0.77	0.65
Radiation element microwave field strength H_{max} (A/m)	4405.7		2569.5
Considering the microwave field strength under the diamond chip model H'_{max} (A/m)	82.8		36.6

For BLA microwave antennas, the overall field strength distribution decreases outward from the center of a small circular notch with a microwave element radius of r (Figure 2) and is symmetrically distributed along the gap g (Figure 2). Therefore, diamond samples should be placed at the center of the small circular notch, and the microwave field strength in the radiation element region at the sample placement is shown in Figure 4a, with a maximum field strength of 82.8A/m. For the DOLR microwave antenna, it is symmetrical along the x-axis at the center (the x-axis and y-axis are in the diamond horizontal plane, and the z-axis is perpendicular to the plane), decreasing along the z-axis. Therefore, the sample should be placed at the center of the antenna, and the field strength near the sample placement is shown in Figure 4b, with a maximum field strength of 36.6 A/m. In terms of microwave field uniformity, we use $(H_{max} - H_{min})/H_{mid}$ to calculate the magnetic field uniformity $H'_{homogeneity}$, and the smaller the value, the better the surface uniformity. For the xy plane, as shown in Table 1, for BLA microwave antennas and DOLR microwave antennas, the smaller the area, the better the uniformity. When the area is less than 1 mm^2, the uniformity is close to but less than 1, and the planar uniformity of the DOLR microwave antenna is slightly better. In addition, the microwave field strength within the range of 0–10 mm from the antenna surface in the z-axis direction has been analyzed. As shown in Figure 5, the field strength of both types of antennas shows a trend of first increasing and then decreasing with increasing distance, whose maximum points are located around 2 mm from the xy plane of the antenna. Therefore, the NV centers are recommended to be placed at about 2 mm away from the surface of the antenna. Moreover, the diamond does not directly contact the antenna, and the heating effect on the diamond is also reduced. Diamond can be placed on 3D printed hollow brackets, or optical tweezers can be used to control the position of the nanodiamonds to achieve this goal. In theory, considering the effects of return loss, thermal effects, magnetic field strength, etc., the DOLR antenna may be more suitable for applications in cellular environments.

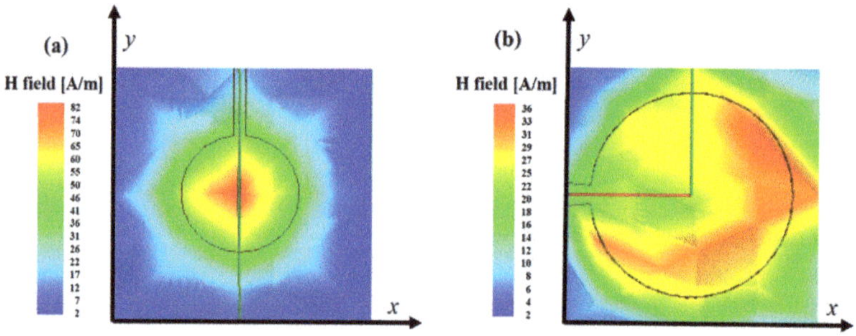

Figure 4. Simulation results of microwave field strength in the radiation element area where diamond samples are placed. (**a**) BLA microwave antenna (H_{max} = 82.8 A/m); (**b**) DOLR microwave antenna (H_{max} = 36.6 A/m).

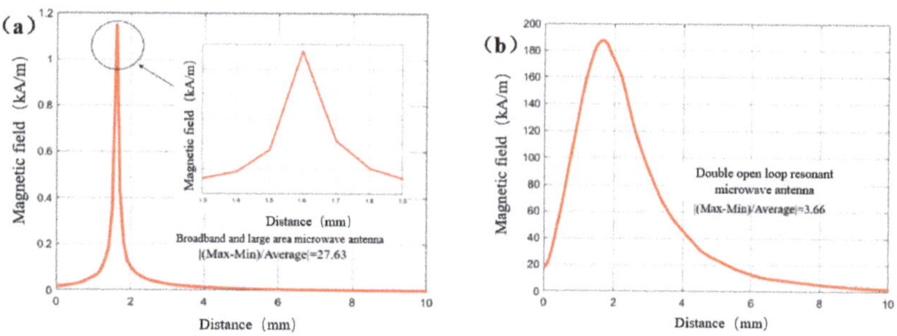

Figure 5. Simulation results of magnetic field strength along z-axis. (**a**) BLA microwave antenna diagram; (**b**) DOLR microwave antenna.

3.2. Model Optimization

In recent years, the *D-T* relationship of NV center thermometers has been studied [37]. For instance, a third-order polynomial has been proposed to fit the *D-T* relationship in the temperature range of 300–600 K [35]. The optimal temperature for human cell activity is known to be (36.5 ± 0.5) °C. If the temperature inside the cells exceeds 42.5 °C or they are irradiated for a long time at this temperature, the active cells, and even tumor cells, will be uniformly killed [38–41]. Therefore, we set the experimental study temperature at 36–42.5 °C, at which $dD/Dt \approx -74.2$ kHz/K, and the corresponding change in D value is about 400 kHz. In the cellular application scenario, diamond does not directly contact the microwave antenna but is swallowed by the cells. The simulations in Figure 5 show that the maximum field intensities of the two irregular antennas are about 2 mm away from the surface, so there is a natural advantage in this aspect. Both cells and diamonds are suspended in a solution environment, so that microwaves need to pass through the culture dish (dielectric constant $\varepsilon_r = 3.01$, loss tangent angle $\tan\delta = 0.0015$), glass slide (dielectric constant $\varepsilon_r = 5.42$, loss tangent angle $\tan\delta = 0.044$), and culture medium (conductivity $\sigma = 5400 \pm 100$ μS/cm) to act on the NV centers. The culture medium, culture dish, and glass slide will affect the center frequency and the heating effect of the microwave antenna. Water is the main component of the culture medium, with a higher specific heat capacity, making it easy to absorb the heat generated by the antenna. Therefore, it directly affects the environmental temperature and cell activity. In the design process, it is necessary to consider correcting the offset of the center frequency in advance and focus on reducing the influence of thermal effects, that is, improving the quality factor Q and $S11$ parameters of the microwave antenna. In addition, media such as culture dishes and culture media can cause microwave power loss. On the premise of ensuring the accuracy of the center frequency and minimal heating effect, the microwave field strength should be increased as much as possible to improve the effectiveness of microwave power acting on cells. Based on the above analysis, we established a model in the cellular environment, as shown in Figure 6.

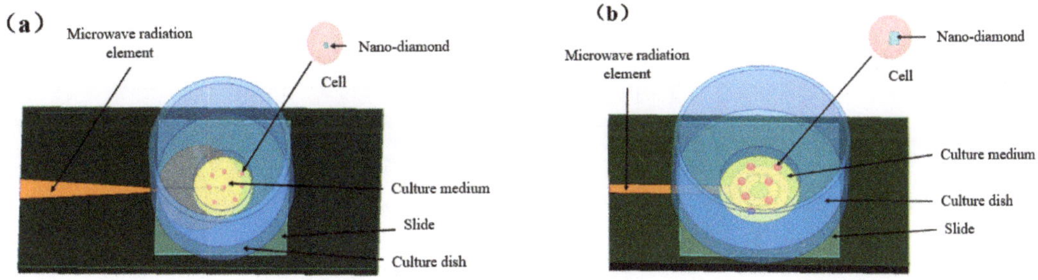

Figure 6. Model diagrams of two types of microwave antennas in a cellular environment. (**a**) Model diagram of the BLA microwave antenna; (**b**) Model diagram of the DOLR microwave antenna.

As shown in Figure 7, materials such as diamond samples and culture dishes will affect the microwave resonance frequency of the antenna, causing a left shift of the center frequency. Compared to the resonance frequency of the original design of the antenna in Table 2, sheet-like diamond only causes a frequency shift of 0.0105 GHz, while the cellular environment causes a frequency shift of 0.4009 GHz for BLA microwave antennas, which is 40 times greater than that in the case of sheet-like diamonds. For the DOLR microwave antenna, the sheet-like diamond causes a frequency shift of 0.0069 GHz, and the cellular environment causes a frequency shift of 0.1589 GHz, which is about 23 times more than that in the case of sheet-like diamonds. The DOLR antenna has a smaller offset than the BLA antennas. For the cellular environment, it is necessary to optimize the antenna size again based on the influence of culture dishes, culture media, and glass slides to reduce the center frequency offset. Moreover, the bandwidth of the BLA antenna can reach 400 MHz, while the bandwidth of the DOLR antenna is only about 20 MHz. It is known that the frequency variation of zero field splitting energy D caused by temperature changes in the cellular environment is about 400 kHz, so both the antennas meet the bandwidth requirements. The heating effect of microwave antennas can be reflected by their quality factor Q value, where the larger the Q value, the smaller the heating effect. As shown in Table 2, the Q value of a BLA antenna is about 7 in different environments. The Q value of a DOLR microwave antenna is about 130 in a single radiation element and diamond sheet, which is about 18 times more than that of BLA antenna. The Q value is about 32 in a cellular environment, which is 4.6 times more than that of a BLA antenna. Therefore, the heating effect of the BLA antenna is larger. Both types of antennas have the highest Q value under the condition of pure radiation element without media. The Q value of the DOLR antenna significantly decreases in the cellular environment, while that of BLA remains almost unchanged.

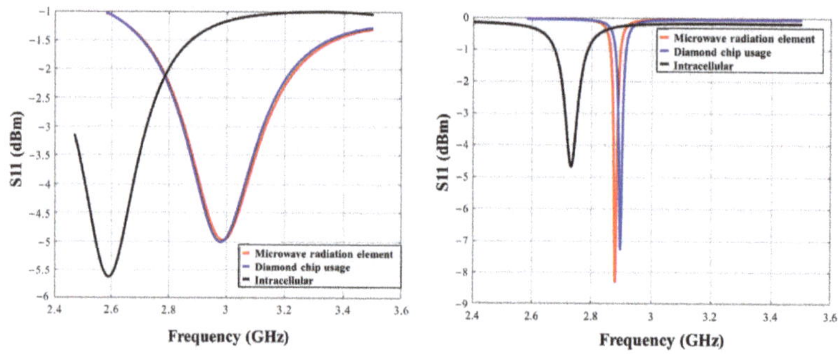

(**a**) (**b**)

Figure 7. Reflection parameter *S*11 of the two microwave antennas under different environmental conditions. (**a**) *S*11 parameters of BLA microwave antenna in different environments; (**b**) *S*11 parameters of DOLR microwave antenna in different environments.

Table 2. The microwave field characteristics of two types of microwave antennas in different environments.

Microwave Field Characteristics \ Antenna Type		BLA Microwave Antenna	DOLR Microwave Antenna
Center frequency (GHz)	Radiation element	2.990	2.890
	Diamond sheet	2.979	2.897
	Cellular environment	2.589	2.731
Bandwidth (MHz)	Radiation element	382	20
	Diamond sheet	422	22
	Cellular environment	337	84
Q value	Radiation element	7.8	148.2
	Diamond sheet	7.1	130.0
	Cellular environment	7.7	32.7
Magnetic field strength (A/m)	Radiation element	2366.5	948.5
	Diamond sheet	2871.5	413.00
	Cellular environment	236.9	161.4
Magnetic field uniformity	Radiation element	2.00	1.93
	Diamond sheet	1.98	1.88
	Cellular environment	0.96	0.42

In terms of magnetic field strength, as shown in Table 2, it can be seen that BLA microwave antennas have a greater advantage in magnetic field strength for diamond sheets, which can still reach over 2000 A/m. However, it drops sharply to 236.9 A/m in cellular application environments. The value of the DOLR microwave antenna is less than 1000 A/m in all the three environments. In the cellular environment, the magnetic field strength of the DOLR antenna is only about 70 A/m less than that of the BLA antenna. In terms of magnetic field uniformity, the diamond sheet environment is analyzed based on a 4 mm $\times$ 4 mm region. The uniformity of the DOLR and BLA antennas is 1.975 and 1.882, respectively. The case of cellular environment is analyzed based on a 1 mm $\times$ 1 mm region. The uniformity of the DOLR and BLA antennas is 0.424 and 0.958, respectively. Thus, the uniformity of the DOLR antenna is better than that of the BLA antenna.

In summary, for the application scenarios of diamond sheets, the advantages of BLA antennas lie in their wide bandwidth and strong magnetic field strength. The advantages of DOLR antennas are low heat loss, high signal-to-noise ratio Q value, and good magnetic field uniformity. For cellular applications, high Q value and low heat loss are required, and DOLR antennas are more suitable for this situation. In addition, it can be concluded that the impact of culture dishes, culture media, and other factors on the antenna in cellular applications is significantly different from that of diamond sheets. Thus, it is necessary to optimize the size of the irregular antennas to make their microwave field characteristics such as center frequency, heat loss, and magnetic field uniformity more in line with the needs of cellular applications.

For cellular application scenarios, the requirements for the center frequency offset, magnetic field uniformity, and heating effects of the microwave field generated by the antenna are different from those of the diamond sheets. The optimization conditions are determined as follows: the center frequency is around 2.87 GHz to achieve the conditions of NV center resonance. Heat loss depletion and return loss should be minimized as much as possible, that is, the Q value and $S11$ parameter should be as large as possible, in order to reduce the impact of heat loss on temperature measurement accuracy and cell activity. Furthermore, the magnetic field uniformity may be improved under the premise of meeting the previous requirements to reduce the impact of NV centers' position on measurement accuracy.

The optimization design of the DOLR antenna is carried out in Figure 8. The original dimensions (as shown in Figure 2b) are as follows. The radii of inner and outer rings are $r_1 = 1.2$ mm, $r_2 = 2.4$ mm and the ring width is $d = r_2 - r_1 = 1.2$ mm. The coupling gap is $g_c = 0.05$ mm, and the gap at the crack is $g_s = 0.4$ mm. The gap between the two rings is $g_r = 0.2$ mm, and the lead width is $w = 1$ mm. Based on the above dimensions, a univariate control method is adopted to simulate the effects of different dimensions on the center frequency and bandwidth. In Figure 8, it can be seen that when the coupling gap g_c is 0.04 mm, both the center frequency and bandwidth can be maximized simultaneously. When the gap between the two rings g_r is 0.2 mm, the center frequency is the highest but the bandwidth is reduced by 20 MHz. When g_r is 0.15 mm, the bandwidth is maximum, but the center frequency is offset by about 0.1 GHz, which is relatively larger compared to the bandwidth offset of diamond itself. The requirement of bandwidth is not high, so priority is given to meeting the center frequency requirement of the ODMR spectral lines. Therefore, a gap g_r of 0.2 mm between the two rings is selected. The inner ring radius r_1 is optimized from 1.0 mm to 1.2 mm. The center frequency is closest to 2.87 GHz with a value of $r_1 = 1.085$ mm. When the lead width w is between 1.2 mm and 1.6 mm, the difference in center frequency and bandwidth is not significant. Considering reducing the thermal contact area and saving metal materials, it is decided to choose a lead width of 1.2 mm. Based on the simulation parameters of $g_c = 0.04$ mm, $g_r = 0.2$ mm, $r_1 = 1.085$ mm, $w = 1.2$ mm, $g_s = 0.4$ mm, and $d = 1.2$ mm, it was found that the center frequency would be larger than 2.87 GHz. Therefore, adjustments are made to r_1 and w in the direction of slightly reducing the center frequency. When the gap g_s is 0.45 mm, the center frequency is maximum, followed by 0.35 mm. In order to bring the center frequency back to around 2.87 GHz, $g_s = 0.35$ mm is chosen. Based on the determination of the first five dimensions, the ring width d ($d = r_2 - r_1$) is adjusted from 1.2 mm to 1.0 mm as shown in Figure 8f. As shown in Figure 9, the center frequency of the antenna is 2.8744 GHz, and the bandwidth is 71 MHz. The return loss $S11$ is -6.8 dBm at the center frequency, and the quality factor Q value is 40.43, which meets the requirements.

Similarly, for BLA microwave antennas, the parameter optimization process is shown in Appendix A. After optimizing the center frequency and bandwidth, the following parameters are selected: $r = 0.20$ mm, $R = 6.98$ mm, $g = 0.2$ mm, and $S = 3.9$ mm. The results are shown by the blue line in Figure 10. At this time, the center frequency is 2.8715 GHz, and the bandwidth is about 311 MHz. The return loss $S11$ is -7 dBm at the center frequency, and the quality factor Q value is 9.22.

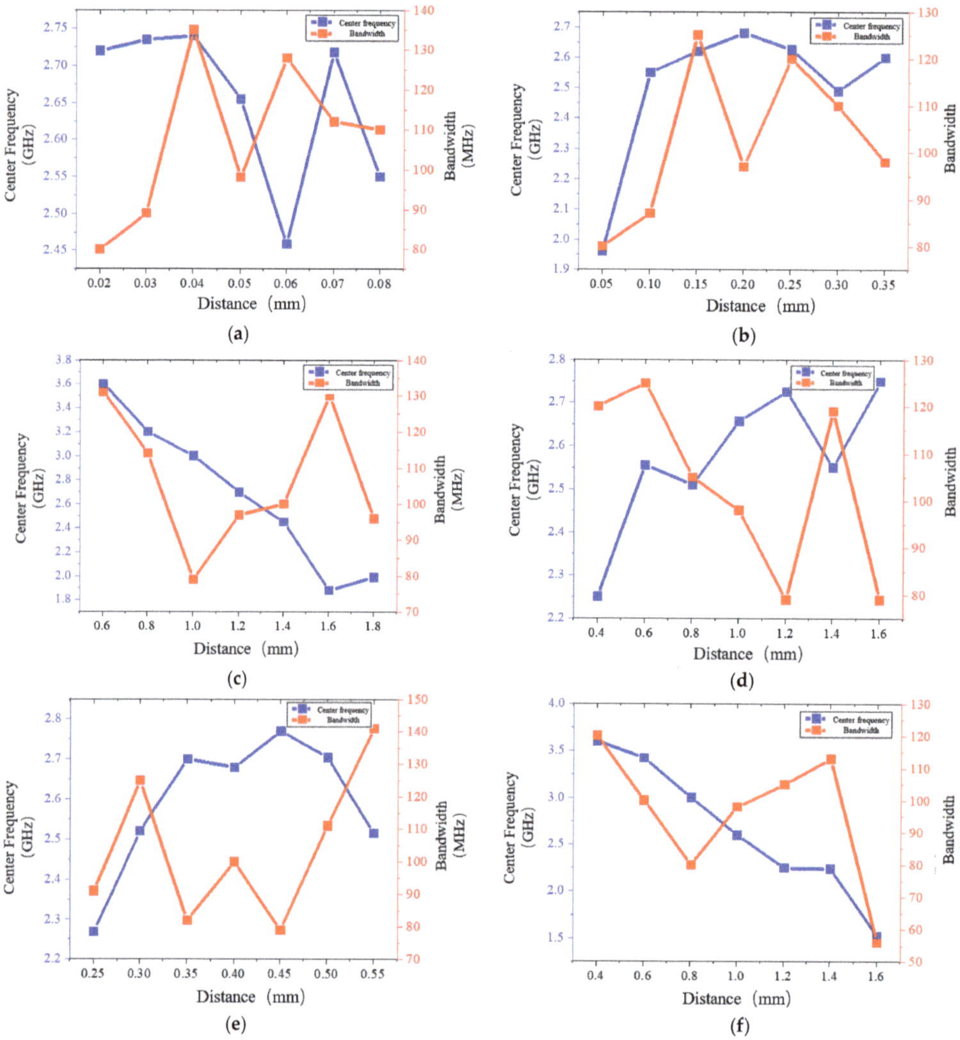

Figure 8. Diagram of the relationship between structural parameters and center frequency of DOLR microwave antennas in cellular application scenarios. (**a**) The relationship between coupling gap g_c and center frequency or bandwidth; (**b**) The relationship between the gap g_r and the center frequency or bandwidth; (**c**) The relationship between inner ring radius r_1 and center frequency or bandwidth; (**d**) The relationship between lead width w and center frequency or bandwidth; (**e**) The relationship between the gap g_s and the center frequency or bandwidth; (**f**) The relationship between ring width d and center frequency or bandwidth.

The optimization results of the above two microwave antennas have been analyzed for Q value, magnetic field uniformity, and magnetic field strength. As shown in Table 3, the Q value of the DOLR antenna is nearly three times that of the BLA antenna. Therefore, the DOLR antenna has smaller heating effect. In terms of planar magnetic field uniformity based on the analysis of 4 mm × 4 mm, that of the BLA antenna is 0.992 larger than that of the DOLR antenna. In terms of magnetic field strength, that of the DOLR antenna is nearly three times that of the BLA microwave antenna. In summary, the DOLR antenna has higher signal-to-noise ratio, better magnetic field performance, and smaller heat effect, making it more suitable for cell research experiments.

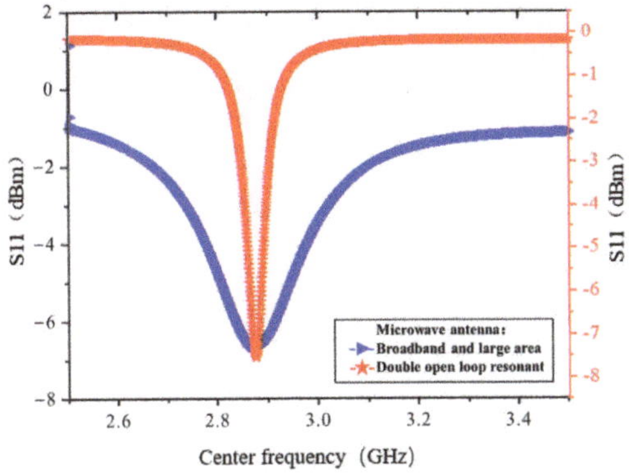

Figure 9. Optimization results of *S*-parameters for BLA microwave antennas (blue line) and DOLR microwave antennas (red line).

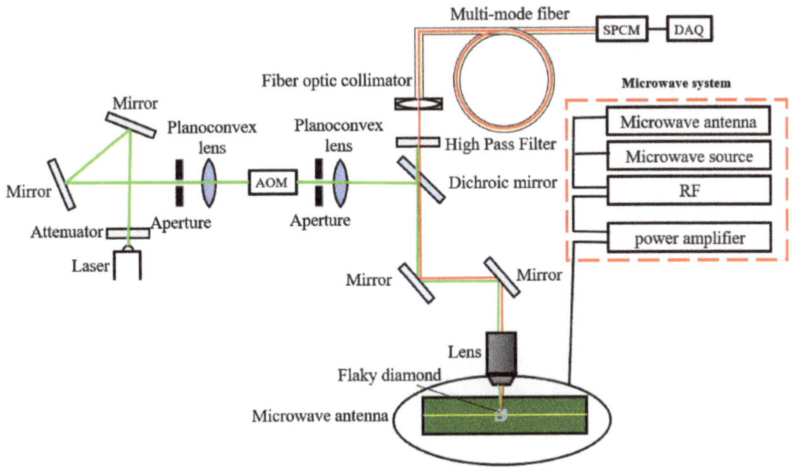

Figure 10. The diagram of ODMR experimental system.

Table 3. Microwave field characteristics of optimized microwave antennas.

Antenna Type	Q Value	Uniformity of Planar Magnetic Field	Magnetic Field Strength (A/m)
BLA microwave antenna	9.22	1.97	144.1
DOLR microwave antenna	40.43	0.98	344.9

4. ODMR Experiments and Discussion

The experimental platform is shown in Figure 10. A continuous laser emitted at 532 nm undergoes collimation and beam expansion after changing the direction of the optical path through planar reflectors. Then, it is incident onto an acousto-optic modulator (AOM) (AODR 1080AF-DIF0-1.0, Gooch & Housego, Ilminster, UK), which generates high-speed switching modulation of the laser. The laser is reflected through a dichroic mirror and enters the objective lens to focus on the sample. The NV centers in diamond emit fluorescence ranging from 600 nm to 800 nm after laser excitation. It passes through a dichroic mirror and the remaining reflected laser at 532 nm passes through a filter (FELH0600, Thorlabs,

Newton, NJ, USA). Finally, the fluorescence is collected by a converging lens and incident onto a single photo counting module (SPCM). A data acquisition (DAQ) system is used to process the number of photons collected by the SPCM. A microwave is generated by a microwave source (SMIQ 06B, Rohde & Schwarz, Columbia, MD, USA), and then amplified by an amplifier (ZHL-16W-43-S+, Mini Circuits, Brooklyn, NY, USA) to a microwave antenna. The RF switch (ZASWA-2-50DRA+, Mini Circuits, USA) is applied to control the on/off of the microwave signal. The system uses a pulse generator (PBESR-PRO-500-PCI, SpinCore, Gainesville, FL, USA) to achieve pulse control of microwave and laser.

Three types of antennas were applied in the experiments. The straight wire microwave antenna used in the comparative experiment has a diameter of 0.06 mm and a length of 3 cm. The dimensions of DOLR and BLA microwave antennas used in the experiment are determined by the optimization in Section 3.2. The ODMR spectral lines of NV centers have been measured at −24 dBm, −20 dBm, −15 dBm, -10 dBm, and −5 dBm based on a straight wire antenna. The changes in contrast and bandwidth are shown in Figure 11. With the increase in microwave power, both the contrast and bandwidth of the ODMR spectrum become larger. The value of zero field splitting energy D shifted to the left, indicating an increase in microwave heating effect and thermal noise. As shown in Figure 12, using line bandwidth/contrast represents the signal-to-noise ratio (SNR) of spectral line measurement. It can be seen that the smaller the value, the higher the SNR and the higher the limit sensitivity level of the system measurement. As the microwave power increases, the SNR first increases and then decreases, indicating the existence of an optimal value. After the microwave power in this system exceeds −10 dBm, the SNR will slightly decrease due to the heating effect and thermal noise. Therefore, when the microwave power is set at −10 dBm, the SNR is optimal, and the heating effect and thermal noise are lower at this time.

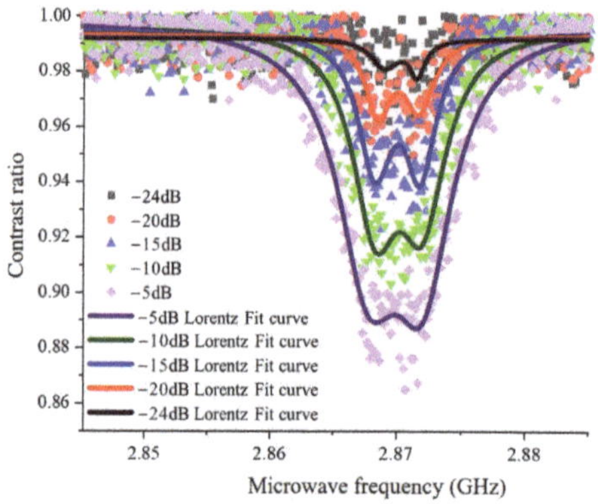

Figure 11. ODMR spectra at microwave fields of −5 dBm, −10 dBm, −15 dBm, −20 dBm, and −24 dBm based on a straight wire antenna.

As shown in Figure 13, the values of bandwidth/contrast for the three antennas have been compared. For DOLR microwave antennas, the SNR reaches its optimal value when the microwave power is around −10 dBm. The optimal microwave power of the straight wire microwave antenna is also around −10 dBm, but its SNR is lower than that of the DOLR microwave antenna. For the BLA microwave antenna, the optimal microwave power is around −15 dBm. The SNR rapidly decreases above −15 dBm, and the SNR remains basically unchanged in the range of −25 dBm to −15 dBm. Its overall SNR is also lower than that of the DOLR antenna. In the range of −25 dBm to −16 dBm, the SNR of the BLA antenna is better than that of straight wire antenna. In summary, within the power range of −25 dBm to −5 dBm, the DOLR antenna has the best ultimate sensitivity and signal-to-noise ratio.

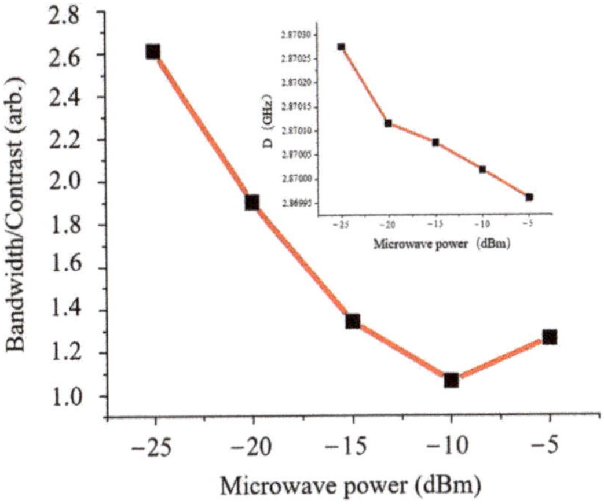

Figure 12. The variation of bandwidth vs. contrast of the measured ODMR spectrum of a straight wire microwave antenna with microwave power.

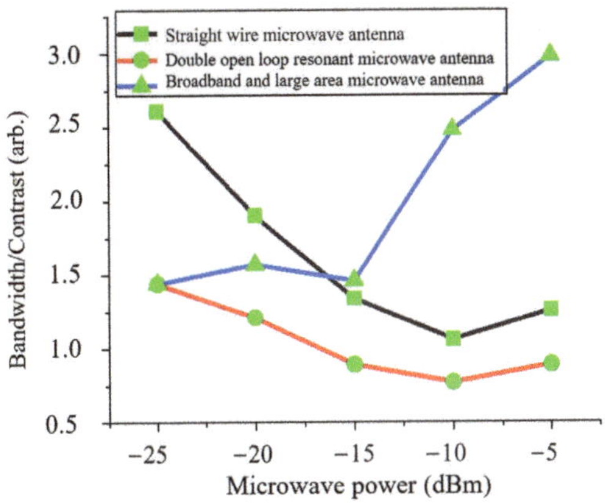

Figure 13. Comparison for bandwidth vs. contrast of the measured ODMR spectrum for the three types of microwave antennas.

It is also important to ensure the accuracy of temperature measurement while improving the signal-to-noise ratio. Therefore, the heating effect of the antenna itself is expected to be as small as possible, and the quality factor Q is increased. The quality factor Q of the three antennas can be calculated by the center frequency/bandwidth (full width at half height), as shown in Figure 14. It can be seen that the DOLR antenna is generally superior to the other two antennas. For the DOLR antenna, the Q value attenuation is not significant in the microwave power range of −25 dBm to −15 dBm. However, when the microwave power is greater than −15 dBm, the Q values will quickly decrease. Therefore, the DOLR antenna has the best SNR at −15 dBm with high quality factor and small heating effect. The Q values of the other two antennas show an approximately linear decreasing trend with increasing microwave power. When the microwave power is less than −15 dBm, the Q value of the BLA antenna is slightly better than that of the straight wire antenna. When the microwave power is greater than −15 dBm, the Q value of the straight wire antenna is higher. Considering both the heating effect and SNR, the DOLR microwave antenna is the optimal choice, and its microwave power can be optimized in the range of −15 dBm to −10 dBm.

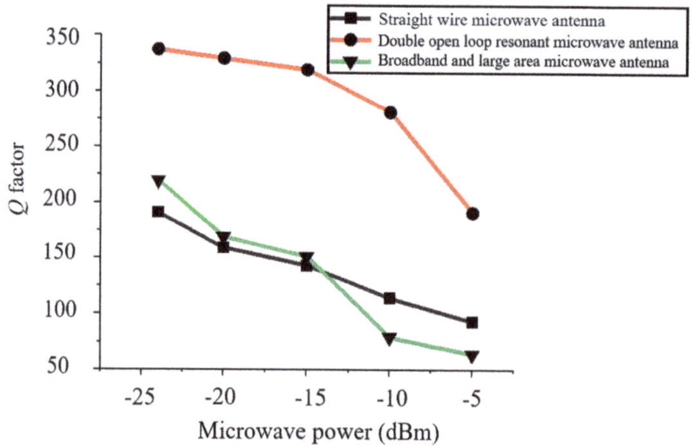

Figure 14. Comparison of *Q* values for the measured ODMR spectrum for the three types of microwave antennas.

5. Conclusions

This paper focuses on the application scenarios of cell temperature measurement, which require microwave antennas with high *Q* value, low return loss, and small heating effect. The comparative analysis was conducted on the characteristics of a DOLR antenna and BLA microwave antenna. The advantages of BLA microwave antennas lie in their wide bandwidth and high magnetic field strength, but their obvious disadvantages lie in their large heating effect and return loss. The advantages of DOLR microwave antennas lie in their high *Q* value, small reflection parameter S11, and the heating effect. Its magnetic field uniformity is good, so the DOLR antenna is more suitable for cellular applications. Moreover, the influence of culture medium, culture dish, etc. on the center frequency, bandwidth, magnetic field properties, etc. of the microwave antennas has been analyzed. The structure parameters of the microwave antenna have been optimized to ensure that the center frequency and bandwidth of the antenna meet the requirements. On this basis, the *Q* value and magnetic field strength should be maximized as much as possible, and the heat loss and thermal effect should be minimized as much as possible. Finally, the signal-to-noise ratio and *Q* value of the straight wire antenna, DOLR antenna, and BLA antenna were quantified and compared through ODMR experiments. The advantages of DOLR microwave antennas of low heating effect and high SNR have been proved. Secondly, the trends of SNR and *Q* value with microwave power variation, as well as the optimal power value, were obtained by power optimization. This article provides a simulation and verification method for designing and optimizing microwave antennas in special environments. Its partial requirements for antennas are also applicable to the magnetic field measurement of NV centers in diamond.

Author Contributions: Investigation, Z.F. and F.W.; Methodology, Z.F. and L.X.; Software, Zhen Fan; Supervision, L.X. and X.F.; Validation, X.F. and J.Z.; Writing—original draft, Z.F. and L.X.; Writing—review and editing, F.W. and X.F. All authors have read and agreed to the published version of the manuscript.

Funding: This work was supported by the Fundamental Research Program of National Institute of Metrology, China (No. AKYZD2209-1) and Fundamental Research Program of National Institute of Metrology, China (No. AKYZZ2207).

Institutional Review Board Statement: Not applicable.

Informed Consent Statement: Not applicable.

Data Availability Statement: The data underlying the results presented in this paper are not publicly available at this time but may be obtained from the authors upon reasonable request.

Acknowledgments: We want to thank all of the agencies mentioned above for their support.

Conflicts of Interest: The authors declare no conflicts of interest.

Appendix A

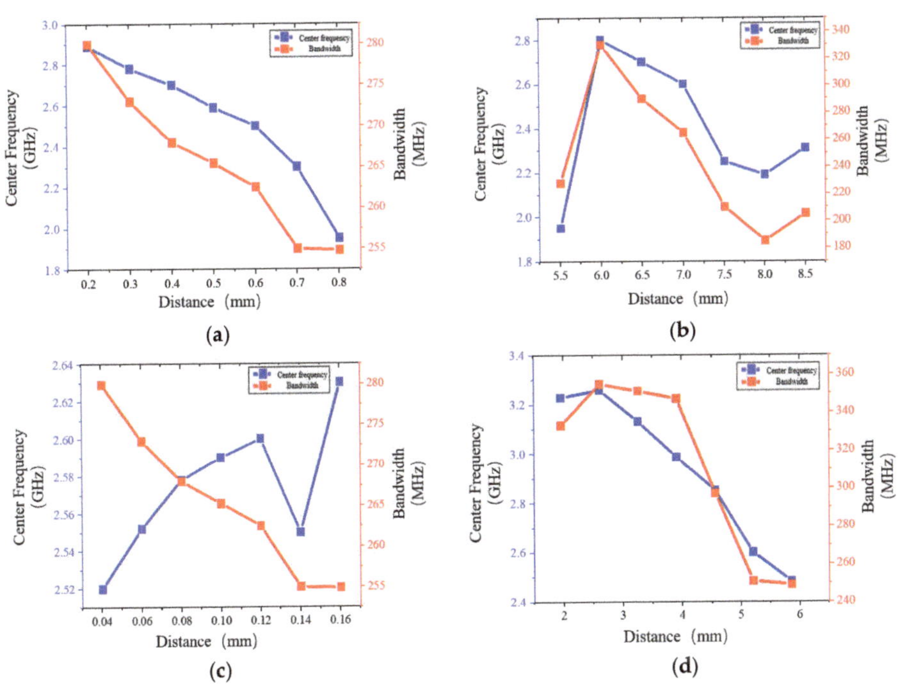

Figure A1. Simulation results of structural parameter optimization of BLA microwave antenna. (**a**) The relationship between ring radius r and center frequency or bandwidth; (**b**) The relationship between hole radius R and center frequency or bandwidth; (**c**) The relationship between gap g and center frequency or bandwidth; (**d**) The relationship between the distance S and center frequency or bandwidth.

References

1. Seymour, R.S. Biophysics and Physiology of Temperature Regulation in Thermogenic Flowers. *Biosci. Rep.* **2001**, *21*, 223–236. [CrossRef] [PubMed]
2. Patel, D.; Franklin, K.A. Temperature-regulation of plant architecture. *Plant. Signal. Behav.* **2009**, *4*, 577–579. [CrossRef] [PubMed]
3. Fan, J.W.; Cojocaru, I.; Becker, J.; Fedotov, I.V.; Alkahtani, M.H.A.; Alajlan, A.; Blakley, S.; Rezaee, M.; Lyamkina, A.; Palyanov, Y.N.; et al. Germanium-vacancy color center in diamond as a temperature sensor. *ACS Photonics* **2018**, *5*, 765–770. [CrossRef]
4. Donner, J.S.; Thompson, S.A.; Kreuzer, M.P.; Baffou, G.; Quidant, R. Mapping intracellular temperature using green fluorescent protein. *Nano Lett.* **2012**, *12*, 2107–2111. [CrossRef] [PubMed]
5. Okabe, K.; Inada, N.; Gota, C.; Harada, Y.; Funatsu, T.; Uchiyama, S. Intracellular temperature mapping with a fluorescent polymeric thermometer and fluorescence lifetime imaging microscopy. *Nat. Commun.* **2012**, *3*, 705. [CrossRef] [PubMed]
6. Ximendes, E.C.; Santos, W.Q.; Rocha, U.; Kagola, U.K.; Sanz-Rodríguez, F.; Fernández, N.; Gouveia-Neto, A.d.S.; Bravo, D.; Domingo, A.M.; Rosal, B.d.; et al. Unveiling in vivo subcutaneous thermal dynamics by infrared luminescent nanothermometers. *Nano Lett.* **2016**, *16*, 1695–1703. [CrossRef] [PubMed]
7. Liu, C.F.; Leong, W.H.; Xia, K.; Feng, X.; Finkler, A.; Denisenko, A.; Wrachtrup, J.; Li, Q.; Liu, R.B. Ultra-sensitive hybrid diamond nanothermometer. *Natl. Sci. Rev.* **2021**, *5*, nwaa194. [CrossRef]
8. Bernardi, E.; Moreva, E.; Traina, P.; Petrini, G.; Tchernij, S.D.; Forneris, J.; Pastuović, Ž.; Degiovanni, I.P.; Olivero, P.; Genovese, M. A biocompatible technique for magnetic field sensing at (sub)cellular scale using nitrogen-vacancy centers. *EPJ Quantum Technol.* **2020**, *7*, 13. [CrossRef]

9. Yu, S.J.; Kang, M.W.; Chang, H.C.; Chen, K.M.; Yu, Y.C. Bright fluorescent nanodiamonds: No photobleaching and low cytotoxicity. *J. Am. Chem. Soc.* **2005**, *127*, 17604–17605. [CrossRef]
10. Fu, C.C.; Lee, H.Y.; Chen, K.; Lim, T.S.; Wu, H.Y.; Lin, P.K.; Wei, P.K.; Tsao, P.H.; Chang, H.C.; Fann, W. Characterization and application of single fluorescent nanodiamonds as cellular biomarkers. *Proc. Natl. Acad. Sci. USA* **2007**, *104*, 727–732. [CrossRef]
11. Xie, Y.; Yu, H.; Zhu, Y.; Qin, X.; Rong, X.; Duan, C.K.; Du, J. A hybrid magnetometer towards femtotesla sensitivity under ambient conditions. *Sci. Bull.* **2021**, *66*, 127–132. [CrossRef]
12. Neumann, P.; Jacobi, I.; Dolde, F.; Burk, C.; Reuter, R.; Waldherr, G.; Honert, J.; Wolf, T.; Brunner, A.; Shim, J.H.; et al. High-precision nanoscale temperature sensing using single defects in diamond. *Nano Lett.* **2013**, *13*, 2738–2742. [CrossRef] [PubMed]
13. Toyli, D.M.; de las Casas, C.F.; Christle, D.J.; Dobrovitski, V.V.; Awschalom, D.D. Fluorescence thermometry enhanced by the quantum coherence of single spins in diamond. *Proc. Natl. Acad. Sci. USA* **2013**, *110*, 8417–8421. [CrossRef] [PubMed]
14. Plakhoynik, T.; Doherty, M.W.; Cole, J.H.; Chapman, R.; Manson, N.B. All-optical thermometry and thermal properties of the optically detected spin resonances of the NV^- Center in nanodiamond. *Nano Lett.* **2014**, *14*, 4989–4996. [CrossRef] [PubMed]
15. Maze, J.R.; Stanwix, P.L.; Hodges, J.S.; Hong, S.; Taylor, J.M.; Cappellaro, P.; Jiang, L.; Gurudev Dutt, M.V.; Togan, E.; Zibrov, A.S.; et al. Nanoscale magnetic sensing with an individual electronic spin in diamond. *Nature* **2008**, *455*, 644–647. [CrossRef] [PubMed]
16. Taylor, J.M.; Cappellaro, P.; Childress, L.; Jiang, L.; Budker, D.; Hemmer, P.R.; Yacoby, A.; Walsworth, R.; Lukin, M.D. High-sensitivity diamond magnetometer with nanoscale resolution. *Nat. Phys.* **2008**, *4*, 810–816. [CrossRef]
17. Ivády, V.; Simon, T.; Maze, J.R.; Abrikosov, I.A.; Gali, A. Pressure and temperature dependence of the zero-field splitting in the ground state of NV centers in diamond: A first-principles study. *Phys. Rev. B* **2014**, *90*, 235205. [CrossRef]
18. Doherty, M.W.; Struzhkin, V.V.; Simpson, D.A.; McGuinness, L.P.; Meng, Y.; Stacey, A.; Karle, T.J.; Hemley, R.J.; Manson, N.B.; Hollenberg, L.C.L.; et al. Electronic properties and metrology applications of the diamond NV− center under pressure. *Phys. Rev. Lett.* **2014**, *112*, 047601. [CrossRef] [PubMed]
19. Dolde, F.; Fedder, H.; Doherty, M.W.; Nöbauer, T.; Rempp, F.; Balasubramanian, G.; Wolf, T.; Reinhard, F.; Hollenberg, L.C.L.; Jelezko, F.; et al. Electric-field sensing using single diamond spins. *Nat. Phys.* **2011**, *7*, 459–463. [CrossRef]
20. Iwasaki, T.; Naruki, W.; Tahara, K.; Makino, T.; Kato, H.; Ogura, M.; Takeuchi, D.; Yamasaki, S.; Hatano, M. Direct nanoscale sensing of the internal electric field in operating semiconductor devices using single electron spins. *ACS Nano* **2017**, *11*, 1238–1245. [CrossRef]
21. Jensen, K.; Leefer, N.; Jarmola, A.; Dumeige, Y.; Acosta, V.M.; Kehayias, P.; Patton, B.; Budker, D. Cavity-enhanced room-temperature magnetometry using absorption by nitrogen-vacancy centers in diamond. *Phys. Rev. Lett.* **2014**, *112*, 160802. [CrossRef] [PubMed]
22. Sasaki, K.; Monnai, Y.; Saijo, S.; Fujita, R.; Watanabe, H.; Ishi-Hayase, J.; Itoh, K.M.; Abe, E. Broadband, large-area microwave antenna for optically detected magnetic resonance of nitrogen-vacancy centers in diamond. *Rev. Sci. Instrum.* **2016**, *87*, 053904. [CrossRef] [PubMed]
23. Bayat, K.; Choy, J.; Farrokh Baroughi, M.; Meesala, S.; Loncar, M. Efficient, Uniform, and large area microwave magnetic coupling to NV centers in diamond using double split-ring resonators. *Nano Lett.* **2014**, *14*, 1208–1213. [CrossRef] [PubMed]
24. Kapitanova, P.; Soshenko, V.V.; Vorobyov, V.V.; Dobrykh, D.; Bolshedvorskii, S.V.; Sorokin, V.N.; Akimov, A.V. 3D uniform manipulation of NV centers in diamond using a dielectric resonator antenna. *JETP Lett.* **2018**, *108*, 588–595. [CrossRef]
25. Qin, L.; Fu, Y.; Zhang, S.; Zhao, J.; Gao, J.; Yuan, H.; Ma, Z.; Shi, Y.; Liu, J. Near-field microwave radiation function on spin assembly of nitrogen vacancy centers in diamond with copper wire and ring microstrip antennas. *Jpn. J. Appl. Phys.* **2018**, *57*, 072201. [CrossRef]
26. Chen, Y.; Guo, H.; Li, W.; Wu, D.; Zhu, Q.; Zhao, B.; Wang, L.; Zhang, Y.; Zhao, R.; Liu, W.; et al. Large-area, tridimensional uniform microwave antenna for quantum sensing based on nitrogen-vacancy centers in diamond. *APEX* **2018**, *11*, 123001. [CrossRef]
27. Fujiwara, M.; Shikano, Y. Diamond quantum thermometry: From fundations to applications. *Nanotechnology* **2021**, *32*, 482002. [CrossRef] [PubMed]
28. Gao, Y.C.; Sun, J.H.; Yu, G.F. Measurement of dielectric constant of cell culture dishes. *J. Univ. Electron. Sci. Technol. China* **2007**, *36*, 70–72.
29. Fujiwara, M.; Sun, S.; Dohms, A.; Nishimura, Y.; Suto, K.; Takezawa, Y.; Oshimi, k.; Zhao, L.; Sadzak, N.; Umehara, Y.; et al. Real-time nanodiamond thermometry probing in vivo thermogenic responses. *Sci. Adv.* **2020**, *6*, eaba9636. [CrossRef]
30. Wang, Z.; Zhang, J.T.; Feng, X.J.; Xing, L. Microwave heating effect on diamond samples of nitrogen-vacancy centers. *ACS Omega* **2022**, *7*, 31538–31543. [CrossRef]
31. Loubser, J.H.N.; van Wyk, J.A. Electron spin resonance in the study of diamond. *Rep. Prog. Phys.* **1978**, *41*, 1201. [CrossRef]
32. Felton, S.; Edmonds, A.M.; Newton, M.E.; Martineau, P.M.; Fisher, D.; Twitchen, D.J.; Baker, J.M. Hyperfine interaction in the ground state of the negatively charged nitrogen vacancy center in diamond. *Phys. Rev. B* **2009**, *79*, 075203. [CrossRef]
33. Barry, J.F.; Turner, M.J.; Schloss, J.M.; Glenn, D.R.; Song, Y.; Lukin, M.D.; Park, H.; Walsworth, R.L. Optical magnetic detection of single-neuron action potentials using quantum defects in diamond. *Proc. Natl. Acad. Sci. USA* **2016**, *113*, 14133–14138. [CrossRef] [PubMed]
34. Schoenfeld, R.S.; Harneit, W. Real time magnetic field sensing and imaging using a single spin in diamond. *Phys. Rev. Lett.* **2011**, *106*, 030802. [CrossRef] [PubMed]

35. Toyli, D.M.; Christle, D.J.; Alkauskas, A.; Buckley, B.B.; Van de Walle, C.G.; Awschalom, D.D. Measurement and control single nitrogen-vacancy center spins above 600K. *Phys. Rev. X* **2012**, *2*, 031001.
36. Doughty, A.C.V.; Hoover, A.R.; Layton, E.; Murray, C.K.; Howard, E.W.; Chen, W.R. Nanomaterial applications in photothermal therapy for cancer. Materials. *Nat. Mater.* **2019**, *12*, 779.
37. Acosta, V.M.; Bauch, E.; Ledbetter, M.P.; Waxman, A.; Bouchard, L.S.; Budker, D. Temperature dependence of the nitrogen-vacancy magnetic resonance in diamond. *Phys. Rev. Lett.* **2010**, *104*, 070801. [CrossRef] [PubMed]
38. Hsiao, C.W.; Chuang, E.Y.; Chen, H.L.; Wan, D.; Korupalli, C.; Liao, Z.X.; Chiu, Y.L.; Chia, W.T.; Lin, K.J.; Sung, H.W. Photothermal tumor ablation in mice with repeated therapy sessions using NIR-absorbing micellar hydrogels formed in situ. *Biomaterials* **2015**, *56*, 26–35. [CrossRef] [PubMed]
39. Chen, Q.; Xu, L.; Liang, C.; Wang, C.; Peng, R.; Liu, Z. Photothermal therapy with immune-adjuvant nanoparticles together with checkpoint blockade for effective cancer immunotherapy. *Nat. Commun.* **2016**, *7*, 13193. [CrossRef]
40. Zhang, C.; Bu, W.; Ni, D.; Zuo, C.; Cheng, C.; Li, Q.; Zhang, L.; Wang, Z.; Shi, J. A polyoxometalate cluster paradigm with self-adaptive electronic structure for acidity/reducibility-specific photothermal conversion. *J. Am. Chem. Soc.* **2016**, *138*, 8156–8164. [CrossRef]
41. Wu, Y.K.; Alam, M.N.A.; Balasubramanian, P.; Ermakova, A.; Fischer, S.; Barth, H.; Wagner, M.; Raabe, M.; Jelezko, F.; Weil, T. Nanodiamond theranostic for light-controlled intracellular heating and nanoscale temperature sensing. *Nano Lett.* **2021**, *21*, 3780–3788. [CrossRef] [PubMed]

photonics MDPI

Article

A Novel Measurement Method for Spin Polarization Three Axis Spatial Distribution in Spin-Exchange Relaxation Free Atomic Magnetometer

Xiujie Fang [1,2], Jin Li [1,2,*], Yanning Ma [1,2,*], Kai Wei [1,2], Wenfeng Fan [1,2], Yueyang Zhai [1,2], Wei Quan [1,2] and Zhisong Xiao [3]

1 Key Laboratory of Ultra-Weak Magnetic Field Measurement Technology, Ministry of Education, School of Instrumentation and Optoelectronic Engineering, Beihang University, Beijing 100191, China
2 Zhejiang Provincial Key Laboratory of Ultra-Weak Magnetic-Field Space and Applied Technology, Hangzhou Innovation Institute, Beihang University, Hangzhou 310051, China
3 School of Physics, Beihang University, Haidian, Beijing 100191, China
* Correspondence: jl11269@buaa.edu.cn (J.L.); by1917071@buaa.edu.cn (Y.M.)

Abstract: The measurement of atomic spin polarization distribution in spin-exchange relaxation free (SERF) magnetometer is an important topic for improving the sensitivity and consistency of multi-channel magnetic field measurement applications. A novel spin polarization spatial distribution measurement method is presented based on the transient response of the magnetometer after modulating the pumped light with a chopper. Polarization is obtained by a slow-down factor based on the fast spin-exchange interaction effects. Longitudinal and transverse polarization distributions are measured simultaneously without interrupting the operation of the SERF status. Under different oscillating magnetic fields, the spin polarization is measured at the cell centroid. Residual magnetic field inside the magnetometer is obtained from the linear relationship between the precession frequency and the oscillating magnetic field. The one-dimensional polarization distributions in the x, y, and z axes are measured using a digital micromirror device with a resolution of 0.25 cm. The measurement results conform to the Lambert-Bier absorption law and the Gaussian distribution law. Furthermore, 7×7 two-dimensional spatial distribution measurements of polarization on the xy and yz planes are performed. Nonuniformity of 1.04 in the xy plane and 1.82 in the yz plane in the built magnetometer. Compared with other measurement methods, the distribution measurement method proposed is independent of optical depth and suitable for low polarization and high polarization applications. Based on the results of the proposed measurement method of spin polarization spatial distribution, further compensation can improve the application consistency of multi-channel magnetic field measurements and improve the sensitivity of single-channel differential measurements.

Keywords: spin polarization spatial distribution; transient response; pump light modulation; the fast spin-exchange interaction effects; spin-exchange relaxation free magnetometers

1. Introduction

Optically pumped alkali metals have been widely used in various areas, such as in atomic magnetometers [1–3], atomic comagnetometers [4,5], atomic clocks [6], nuclear magnetic resonance [7,8], quantum memory, and teleportation [9,10]. In the field of atomic magnetometer, the spin-exchange relaxation free (SERF) magnetometer achieved sub-fT level sensitivity due to the elimination of spin exchange relaxation between atoms [11]. The SERF magnetometer has characteristics such as low cost, miniaturization, and integration, with significant application potential in biological measurement fields, such as magnetoencephalography (MEG) and magnetocardiography (MCG) measurements [12–14]. The spin polarization of alkali metal atoms is a key factor affecting the performance of the SERF magnetometer systems [1]. The inhomogeneity of atomic polarization affects the imaging accuracy of MEG and MCG, and the polarization gradient limits the further improvement of the sensitivity of SERF magnetometer [15].

The spin polarization of atoms has a nonuniform distribution in space with the distribution of the light field. Realizing the SERF state requires conditions of extremely weak magnetic fields and high temperatures. The SERF magnetometer operates at a high atomic density of 10^{13}–10^{14} cm^3, resulting in a large optical depth in the direction of the pumped light [16]. The nonuniform distribution of polarizability in the longitudinal direction is due to the large optical depth that causes the light intensity to decay quickly [17]. In addition, the pumped light produces a transverse nonuniform polarization distribution for a Gaussian-distributed light field in the transverse direction [18]. The spatial distribution of polarization, rather than its mean value, is more important to us. The measurement of the spatial distribution of spin polarization is highly significant for achieving high-precision and high spatial resolution magnetic field measurement.

Currently, alkali atom polarization measurement methods mainly include the electron paramagnetic resonance (EPR) method, radio frequency spectroscopy (RFS) method, and pumping decay transient (PDT) method. Appelt et al. [19] used the EPR method to measure a two-dimensional spin polarization distribution. Young et al. [20] used the RFS method to apply a transverse RF field under a large background magnetic field and obtained the two-dimensional distribution of atomic electron spin polarizability in the *x*–*z* plane with spatial resolution less than 0.25 cm under the condition of spatial position coding using magnetic field gradient. However, the normal operation of the SERF magnetometer was affected by large magnetic and radiofrequency fields. Walker et al. [21] measured the pumping and relaxation rate using a chopper to rapidly turn on and off the pumping light and determine the polarization of alkali metal atoms. Gusarov et al. [22] constructed an early model of on–off pumping light for the SERF magnetometer at very low polarizability with a constant slowing-down factor for scalar and vectorial magnetic field measurements. The slowing-down factor in the spatial spin polarization distribution of SERF systems could not be regarded as a fixed value. Li et al. [23] used the transmitted intensity attenuated by the pumping light to obtain the average initial intensity of the SERF magnetometer. When the optical depth was too large, the pumped light could hardly pass through the cell and the method could not measure the spatial distribution. In addition, the vast majority of the measured polarization distribution methods were longitudinal planes in the pumping light direction, with no transverse-plane-polarization distribution measurement method.

This study proposes a novel method for measuring alkali atoms' spatial polarization distributions in a SERF magnetometer. Based on the fast spin-exchange interactions, the transverse and longitudinal plane distributions of alkali metal polarization can be obtained from the transient response of the magnetometer after modulating the pumped light using a chopper. Pumped light fully polarized alkali metal atoms in an open state and completely depolarized alkali metal in a closed state. This method is advantageous because it can measure the spin polarization spatial distribution in the SERF regime, and it is not affected by the optical depth and does not require the application of large background and radiofrequency fields. The method is suitable for both low and high polarization.

In addition, we used a digital micromirror device (DMD) for distribution measurements to acquire precise high spatial resolution spatial polarization distributions, which we have validated as a feasible high spatial resolution detection method [24]. Unlike the methods of moving the photodetector (PD) using a displacement console [25] and a large reflector [26], this method avoids variability in the measurement process and problems with the inability to precisely locate it, as well as avoids smear and blooming effects caused by the use of a charge-coupled device [27]. We used a square vapor cell in the experiments to avoid measurement error due to optical distortion caused by the vapor cell curvature. The resolution of spin polarization spatial distribution was 0.25 cm. The experimental results were consistent with the theoretical calculations and simulation results, demonstrating the validity and practicability of the proposed measurement method. The spin polarization spatial distribution measurement method optimized the performance of SERF magnetometers, and, as such, the application consistency of the multichannel measurement of the magnetic field can be improved after taking means of compensation. Our proposed measurement method is applicable to other types of optical pumping atomic magnetometers.

2. Basic Principles

2.1. Principle of Slowing-Down Effect in the SERF Regime

Spin-exchange collisions cause atomic spin relaxation. In traditional magnetometers, the accuracy of magnetometers is often limited by the relaxation caused by spin-exchange collisions. [28]. Under conditions of high density and low magnetic field, the spin-exchange collision relaxation rate between alkali metal atoms can be significantly attenuated to achieve a SERF state [29]. The spin-exchange collision rate between atomic spins is generally 10^5–10^6 s^{-1} in magnitude, which is much larger than the optical pumping and relaxation rates. Atomic spin obeys the spin temperature distribution between superfine Zeeman levels [30]. The spin temperature parameter β is related to the atomic polarization as follows [17]:

$$\beta = \ln\left(\frac{1+P_z^e}{1-P_z^e}\right). \tag{1}$$

The probability density is $\rho \propto e^{\beta F_z}$. F_z is the angular momentum size in the z direction of the Zeeman Sublevel, and P_z is the polarization in the z direction.

Atoms at two superfine levels, $F_a = I + 1/2$ and $F_b = I - 1/2$, obey the spin temperature distribution at each Zeeman level under the condition of fast spin-exchange collisions. In an external magnetic field along the y direction, the momentum causing the precession is mainly determined by the electron spin. The nuclear spin of an alkali metal atom and the electron spin coupling slows the frequency of the electron spin. The equivalent spin-to-magnetic ratio γ_{equ} that decreases due to the existence of the nuclear spin is as follows [31]:

$$\gamma_{equ} = \pm\frac{\gamma^e}{2I+1}, \tag{2}$$

where the gyromagnetic ratio of electrons is $\gamma^e \approx 2\,\pi \times 28$ Hz/nT, and I is the nuclear spin quantum number, with different values for different types of alkali metals.

The sum of the mean total angular momentum of the spin of an atom at the superfine level of the ground state is $\vec{F}_a + \vec{F}_b$ in the z direction. After the moment Δt, the counter-clockwise rotation angle of an atom in a state a is $\omega_0\Delta t$, atoms in state b rotate the same angle clockwise $\omega_0\Delta t$, and their sum of total angular momentum is $\vec{F}_a{}' + \vec{F}_b{}'$. Under SERF conditions, the Lamorr precession frequency of the atomic spin is significantly lesser than the spin-exchange collision rate between alkali metal atoms. According to the principle of conservation of angular momentum, the formula for the conservation of system angular momentum in the x direction is as follows:

$$-\,\omega_0\Delta t F_a + \omega_0\Delta t F_b = \omega\Delta t(F_a + F_b), \tag{3}$$

where ω_0 is the Larmor precession frequency of atomic spin in the external magnetic field under the condition of complete polarization, and ω is the precession frequency of the system's total angular momentum. Combining with the above, m_F is the angular quantum number of an alkali atom at the hyfine level F. The overall precession frequency is given as follows:

$$\omega = \frac{-\sum\limits_{m_F=-(2I+1)}^{m_F=2I+1} m_F e^{\beta m_F} + \sum\limits_{m_F=-(2I-1)}^{m_F=2I-1} m_F e^{\beta m_F}}{\sum\limits_{m_F=-(2I+1)}^{m_F=2I+1} m_F e^{\beta m_F} + \sum\limits_{m_F=-(2I-1)}^{m_F=2I-1} m_F e^{\beta m_F}}\,\omega_0. \tag{4}$$

The precession frequency of the system is related to the spin temperature distribution parameter β, which is related to the atom's spin polarization, according to Equation (1). According to the spin temperature distribution [32], (4) can be expressed as follows:

$$\omega = \frac{(2I+1)\omega_0}{q(P)} = \frac{\gamma_e B}{q(P)}. \tag{5}$$

The slow-down factor $q(P)$ slows the precession frequency of the alkali metal atoms in the SERF regime. The ability to determine whether the precession frequency of the alkali metal atoms under the external magnetic field is further slowed based on the Larmor precession frequency is crucial in determining whether the atoms are in the SERF working state. $q(P)$ is determined by the nuclear spin quantum number I and the spin polarization P. Table 1 shows the commonly used expressions of $q(P)$ for alkali metal atoms. The $q(P)$ is regarded as a function of spin polarization P in the accurate measurement of atomic spin polarization.

Table 1. Nuclear spin Quantum Number and Slow-down Factor of Alkali Metal Atoms.

Alkali Metal Atoms	Nuclear Quantum Number I	Natural Abundance	Slow-Down Factor $q(P)$	Range of Slow-Down Factor $q(P = 0\sim1)$
^{85}Rb	5/2	72.2%	$\frac{38+52P^2+6P^4}{3+10P^2+3P^4}$	38/3~6
^{87}Rb	3/2	27.8%	$\frac{6+2P^2}{1+P^2}$	6~4
^{39}K	3/2	93.3%	$\frac{6+2P^2}{1+P^2}$	6~4
^{41}K	3/2	6.7%	$\frac{6+2P^2}{1+P^2}$	6~4
^{133}Cs	7/2	100%	$\frac{22+70P^2+34P^4+2P^6}{1+7P^2+7P^4+P^6}$	22~8

2.2. Theoretical Model of Spin Polarization Measurement

Under rapid spin-exchange conditions, the spin motion of alkali metal atoms can be described quantitatively by the evolution of the density matrix. The dynamic equation of the interaction between atomic spin and external magnetic field can also be described using a more intuitive and simple Bloch equation model [17]. If the x axis represents the direction of detection light and the z axis represents the direction of pumping light, the matrix form of the Bloch equation can be expressed as follows:

$$\frac{d}{dt}\mathbf{P} = \hat{G}\mathbf{P} + \begin{pmatrix} 0 & 0 & R_{op} \end{pmatrix}^{\mathrm{T}}, \tag{6}$$

where $\mathbf{P} = \left[P_x, P_y, P_z\right]^T$ is the spin polarizability vector, and $\hat{G}$ is the interaction matrix:

$$\hat{G} = \begin{bmatrix} -\frac{R_{tot}}{q(P)} & -\gamma^e \frac{B_z}{q(P)} & \gamma^e \frac{B_y}{q(P)} \\ \gamma^e \frac{B_z}{q(P)} & -\frac{R_{tot}}{q(P)} & -\gamma^e \frac{B_x}{q(P)} \\ -\gamma^e \frac{B_y}{q(P)} & \gamma^e \frac{B_x}{q(P)} & -\frac{R_{tot}}{q(P)} \end{bmatrix}, \tag{7}$$

where B_x, B_y, and B_z are the magnitudes of the magnetic field applied by each of the three axes perceived by the alkali metal atoms; $R_{tot} = R_{op} + R_{rel}$, and R_{op} are the pumping rates; and R_{rel} represents the total relaxation rate, excluding the relaxation effect caused by pumping light. The SERF regime, effectively, inhibits spin-exchange relaxation, with spin destruction relaxation being the main relaxation term. The spin destruction relaxation mainly comes from the binary collision between an alkali metal atom and a buffer gas He atom, the binary collision between an alkali metal atom and a quenched-gas N_2 molecule, the binary collision between alkali metal atoms, and the collision relaxation between alkali metal atoms and the vapor cell wall.

In this study, we use a chopper to make square wave amplitude modulation of the pumping light at a 50% duty cycle; the pumping rate corresponding to the device's initial pumping rate is R_P, ignoring the higher-order harmonic retaining direct current and first harmonic, and the pumping rate can be written as follows:

$$R_{op} = \frac{R_p}{2}\left[1 + \frac{4}{\pi}\cos(\omega_{\text{mod}}t)\right], \tag{8}$$

when there is only a magnetic field B_y, and the pumping rate R_{op} is less than the relaxation rate R_{rel}. The solution of the Bloch Equation (6) above is given as follows [33]:

$$\begin{aligned} P_x(t) &= R_p[T_1(t) + S_1(t)] \\ P_z(t) &= R_p[T_2(t) + S_2(t)] \end{aligned} \tag{9}$$

The first term $T(t)$ is the damping attenuation term in the dynamic solution:

$$\begin{aligned} T_1(t) &= e^{-\left(\frac{R_{rel}t}{q(P_z^0)}\right)} R_\text{p}\left[(k_0+k_1)\omega_y\cos\left(\frac{\omega_y t}{q(P_z^0)}\right) + (k_0+k_1)R_{rel}\sin\left(\frac{\omega_y t}{q(P_z^0)}\right)\right], \\ T_2(t) &= e^{-\left(\frac{R_{rel}t}{q(P_z^0)}\right)} R_\text{p}\left[(k_0+k_1)R_{rel}\cos\left(\frac{\omega_y t}{q(P_z^0)}\right) - (k_0+k_1)\omega_y\sin\left(\frac{\omega_y t}{q(P_z^0)}\right)\right], \end{aligned} \tag{10}$$

and the steady-state oscillation term $S(t)$ is as follows:

$$\begin{aligned} S_1(t) &= \omega_y R_\text{p}\begin{pmatrix} -k_0 + 2k_2\left(\omega_y{}^2 + R_{rel}{}^2 - q(P_z^0)^2\omega_{\text{mod}}{}^2\right)\cos(\omega_{\text{mod}}t) \\ +4k_2 q(P_z^0)\omega_{\text{mod}}R_{rel}\sin(\omega_{\text{mod}}t) \end{pmatrix}, \\ S_2(t) &= R_\text{p}\begin{pmatrix} -k_0R_{rel} + 2k_2R_{rel}\left(\omega_y{}^2 + R_{rel}{}^2 + q(P_z^0)^2\omega_{\text{mod}}{}^2\right)\cos(\omega_{\text{mod}}t) \\ +2k_2 q(P_z^0)\omega_{\text{mod}}\left(-\omega_y{}^2 + R_{rel}{}^2 + q(P_z^0)^2\omega_{\text{mod}}{}^2\right)\sin(\omega_{\text{mod}}t) \end{pmatrix}. \end{aligned} \tag{11}$$

where k_0, k_1, and k_2 are given as follows:

$$\begin{aligned} k_0 &= -\frac{1}{2\left(R_{rel}^2+\omega_y{}^2\right)}, \\ k_1 &= -\frac{1}{\pi\left(R_{rel}^2+\left(\omega_y-q(P_z^0)\omega_{\text{mod}}\right)^2\right)} - \frac{q(P_z^0)+\omega_y}{\pi\left(R_{rel}^2+\left(\omega_y+q(P_z^0)\omega_{\text{mod}}\right)^2\right)}, \\ k_2 &= \frac{1}{\pi\left(R_{rel}^2+\left(\omega_y-q(P_z^0)\omega_{\text{mod}}\right)^2\right)\left(R_{rel}^2+\left(\omega_y+q(P_z^0)\omega_{\text{mod}}\right)^2\right)}. \end{aligned} \tag{12}$$

A circularly polarized pumping laser is used to polarize alkali metal atoms, and another linearly polarized detuning laser is used to detect magnetic field signals in a dual-beam magnetometer. The optical eigenmode of linearly polarized light is the superposition of left and right circularly polarized light. Linearly polarized light shows circular birefringence through the polarized vapor cell, and the rotation angle is produced due to the Faraday rotation effect. When the detected light direction is along the x direction, the rotation angle θ is [34]:

$$\theta = -\frac{\pi}{2}nlr_e cPf_{D2}\frac{v_{\text{pr}} - v_{D2}}{\left(v_{\text{pr}} - v_{D2}\right)^2 - \left(\Gamma_{D2}/2\right)^2}, \tag{13}$$

where l is the distance at which the probe light interacts with alkali metal atoms, n is the rubidium atom's density number, c is the propagation speed of light, r_e is the electron radius, and v_{pr} is the frequency of the probe light. Γ_{D2}, f_{D2}, and v_{D2} are the pressure broadening, oscillation intensity, and resonance frequency of the probe light in the Rb D2 line.

To suppress low-frequency noise and improve detection sensitivity, a photo-elastic modulator (PEM) is used to detect the optical rotation angle, and a PD is used to convert optical signals into electrical signals. The final output V_{out} of the device is as follows:

$$V_{out} = \eta I_0 \alpha e^{-OD(v)} \theta, \quad (14)$$

where η is the PD's conversion coefficient, I_0 is the incident probe light intensity, α is the modulation angle of PEM, and OD is the optical depth.

Figure 1 shows that when light is probed along the x direction, the output voltage V_x is proportional to P_x, V_z is proportional to P_z, and the output signal records the oscillation and steady-state processes of P_x and P_z according to Equations (9), (10), and (12); through the oscillation process, we can obtain the slow-down factor $q(P)$ from the oscillation frequency $\omega_{of} = \omega_y / q(P)$. Figure 2 shows the trajectory of atomic spin polarization with time inside the cell. The oscillation frequency of the output signal P_z can be used to obtain $q(P)$ while probing light along the z direction. Table 1 shows that the value of the initial spin polarization P is calculated from the relationship between $q(P)$ and P. We measure the spatial distribution of the spin polarization using the DMD to scan the x and z direction probe light to obtain the atomic polarization distribution in the longitudinal direction (yz-plane) and the transverse direction (xy plane) of the device, respectively.

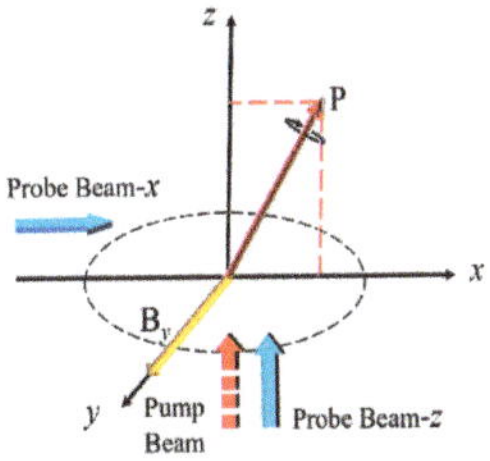

Figure 1. When the magnetic field is input along the y axis, the spin polarization of the atoms deviates from the pumping z axis and stabilizes at the angle between the z axis and the x axis; the x and z axis probe light undergoing an oscillation and decay process under the action of the pumping light modulation probes the x and z axis polarization responses, respectively.

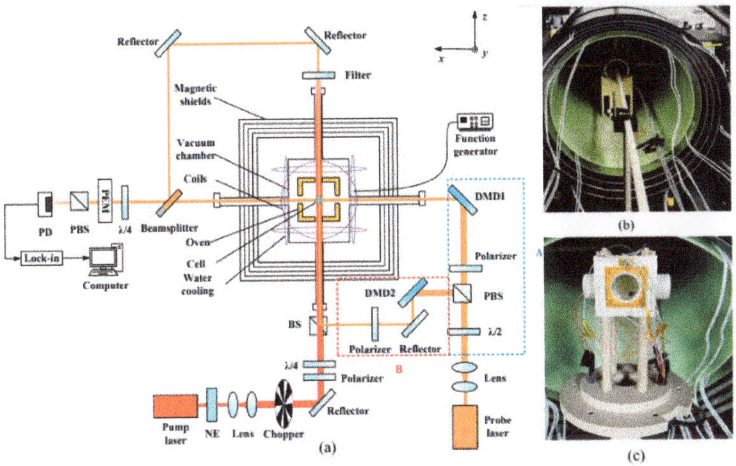

Figure 2. (**a**) Experimental diagram of the SERF magnetometer. NE: noise eater; PEM: photo elastic modulator; GT: Glan–Taylor polarizer; PD: photodetector. (**b**) Photographs of the five-layer magnetic shielding and nonmagnetic vacuum system. (**c**) Photographs of the nonmagnetic electric heating system and the square cell in the center.

3. Experimental Setup

The experimental measurement setup shown in Figure 2 is a dual-beam Rb atomic magnetometer in the SERF regime. The key component of the magnetometer was an aluminosilicate glass (GE180) 2.5-cm-length cubic vapor cell containing 50 Torr quenching gas N_2 and 340 Torr buffer gas ^4He in a liquid nitrogen environment and Rb alkali metals in natural abundance. The cell was placed in a boron nitride ceramic oven in the center of the device. The cell was heated with a nonmagnetic twisted heating wire with 75 kHz AC and real-time temperature monitoring using platinum resistors. The magnetometer had a vacuum inside it to maintain temperature stability, and the vacuum assembly was made of polyether ether ketone material. Laser absorption spectroscopy was used to determine atomic density at high temperatures, which was then calculated using Raoul's law and the saturation vapor density equation. The magnetic shielding system was made up of five layers of stacked cylindrical high-permeability μ-metals [35]. A three-axis compensation coil was configured to further ensure an extremely weak magnetic environment. The magnetic coils were driven by two high-precision function generators (33500 B Keysight, the USA). This ensured that the atoms were in a SERF state.

The alkali atom was polarized by the distributed Bragg reflector (DBR) diode-pump-laser along the z axis. The pump light was 2.5 cm circularly polarized after a lens beam expansion system and a quarter-wave plate, and its wavelength tuned to the D1 resonance line of Rb. A noise eater (Thorlabs, NEM03L) was used to stabilize a laser power of 13.5 mW. The pumped light was modulated using a chopper. The probe light was generated by another DBR diode laser with a wavelength detuned by 1 nm over the D2 resonance line of Rb. The linearly polarized light of 2 cm after PBS was divided into two probe beams. One beam was used to detect the transverse polarization component P_x along the x axis, as shown in part A of Figure 2, and the other beam was used to detect the longitudinal polarization component P_z along the z axis after beam splitter, as shown in part B. Only probe light passed through the filter at the barrel's z axis exit. DMD (Texas Instruments V-9501) with a pixel size of $10.8 \times 10.8\ \mu m^2$ was used in our experiment to obtain the spatial distribution of polarization with high spatial resolution. The grayscale image loaded onto the DMD was used to control the probe light reflected onto the DMD. In the grayscale image, a pixel gray value of 0 indicated the on-state, whereas 256 indicated the off state. Only the probe light in the on-state could pass through the cell to detect the response signal in the spatial position. The probe light beam was divided into several units by DMD, each of which contained 232 pixels with an area of $0.25 \times 0.25\ cm^2$, with a detected optical power of 3 mW per unit. Experiments were conducted to measure the spatial distribution of longitudinal and transverse polarization in 7×7 units. The DMD and the cubic cell were precisely positioned to correlate to ensure a clear measurement position [24]. To reduce the measurement error, the two probe beams passed through the same PEM (Model-100, Hinds Instruments) and PD (Thorlabs, PDA36A2) after passing through the semitransparent and semi-reflective beam splitter. The output signal was transmitted to a phase-locked amplifier ZI (Zurich Instruments HF2LI) for demodulation.

In the experiment, there were two processes of pumping light on and off when a chopper was used to modulate the pumping light. The process of pumping light on required the consideration of both the optical pumping rate R_{op}, the atomic relaxation rate R_{tot}, and the pumping light off state $R_{op} = 0$. We simulated the aforementioned two processes under different B_y magnetic fields with time based on the solution of Equation (6) in continuous pump light. The experimental temperature was 150 °C. The atomic density $n = 8.91 \times 10^{13}\ cm^{-3}$ and the pumping rate $R_{op} = 95.66\ s^{-1}$ at the center of the vapor cell. Figure 3a shows the simulation of the pumped light on-state, and Figure 3b shows the simulation of the pumped light off state. Oscillatory processes of spin polarization increased as the B_y magnetic field increased. The oscillatory process of the polarization was more pronounced in the pumped light off state and was more beneficial for fitting and measuring our data. We also observed the polarization decay time during the pumping light off period to determine the frequency of chopper modulation of pump light. The atoms must be fully depolarized within the modulation period to ensure measurement accuracy. The spin polarization decay times to zero (~10^{-6}) were 0.542, 0.52, 0.497, and 0.47 s when B_y was 1 nT, 4 nT, 7 nT, and 10 nT, respectively. This shows that the oscillatory process intensifies, as well as the decay process, when B_y increases. In addition, considering the requirement for fast measurements, the pump laser beam was modulated by a chopper with a duty cycle of 50% at a rate of 0.8 Hz.

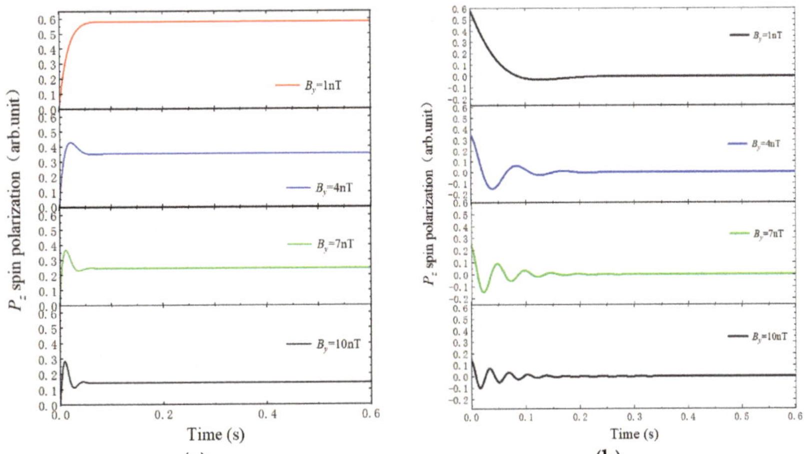

Figure 3. Simulation results of the magnetometer z axis polarization oscillation decay process at different magnetic fields B_y. (**a**) Pumping light in the on state. (**b**) Pumping light in the off state.

The SERF magnetometer polarizability spatial distribution measurement procedure was as follows: First, the compensation coils compensated for the residual magnetic field after demagnetization, such that the residual magnetic field in the magnetometer was approximately zero. y axis coils generated a magnetic field B_y and turned on the chopper. The first gray image on the DMDs was loaded and selected the first measurement positions. The part A probe light path in Figure 2 was turned on, and the part B probe light path was turned off to measure the spatial polarization distribution of the yz plane through the P_x signal response. The part B probe light path was turned on, and the part A probe light path was turned off to measure spatial polarization distribution of the xy plane through the P_z signal response. Signal response data were acquired using LabVIEW and stored in an NI (PXIe-1092PXIe, National Instruments) system. The periodic signal was extracted and fitted to Equation (10), and $q_{xy}(P)$ and $q_{yz}(P)$ were calculated from the fitting results. Therefore, P_{yz} and P_{xy} could be calculated according to Table 1. The grayscale image at the next spatial location was replaced to measure the next point of polarization, resulting in a two-plane 7×7 polarization spatial distribution measurement.

4. Results and Discussion

4.1. Polarization Measurements Based on Pump Light Modulation

To investigate the practicability of the proposed method, the central point was used to measure the polarization by the pump light modulation on the transient response. Table 2 describes the experimental conditions. Figure 4a,b show the experimental measurement data using path A (P_x transient response) and path B (P_z transient response) at B_y = 7 nT. The transient response signal of the magnetometer was fitted with oscillation attenuation in one cycle using Equation (10). The experimental values fit well with the theoretical equations, with an R-square value of 0.9997 for two fitting curves, indicating that the method based on the transient response of the magnetometer was feasible. Tables 3 and 4 show the oscillation frequency ω_{of} measured under different B_y. Then, we calculated the slow-down factor $q(P)$ and precession frequency ω from B_y and ω_{of}. The inset of Figure 4a,b shows the linear relationship between ω and B_y. The linear fit intercept is the residual magnetic field inside the magnetometer. The residual magnetic field B_{res} was 1.27 nT according to the fitting results, which included the residual magnetic field after the demagnetization of the magnetic shield caused by light shifts [36].

Table 2. Parameters of Measurement System.

Parameter	Value	Unit
Diameter of the cubic cell	2.5	cm
Coil constant of z axis	114.03	nT/V
Coil constant of y axis	15.23	nT/V
Coil constant of x axis	15.24	nT/V
Wavelength of the pump laser	794.972	nm
Wavelength of the probe laser	779.242	nm
Power of the pump laser	23	mW
Power of the probe laser (each point)	15	mW
Modulation frequency of PEM	50	kHz
Modulation amplitude of PEM	0.08	rad
Temperature	150	°C

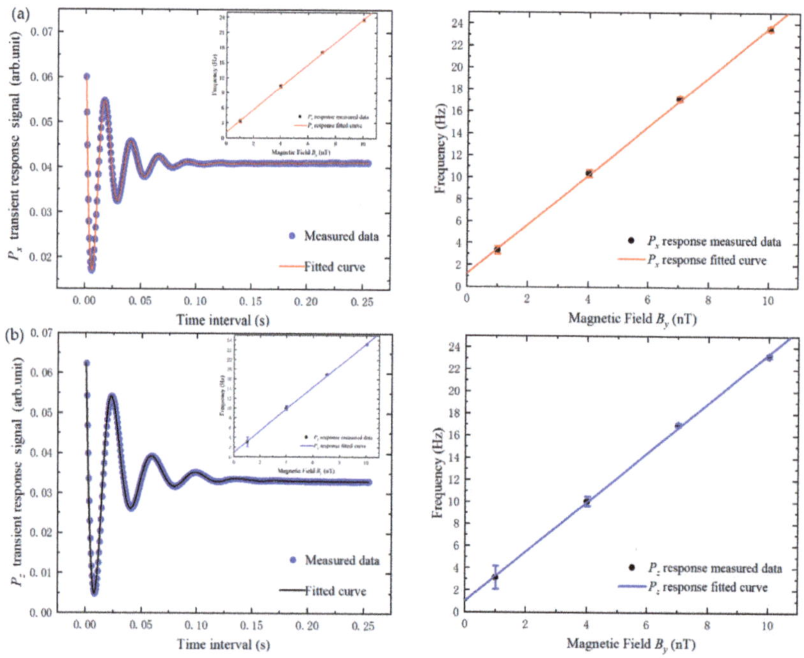

Figure 4. Experimental measurement data of the magnetometer transient response at B_y = 7 nT. The dots are the measured data, and the solid line is the curve fitted to Equation (10). The inset measures the variation in ω with B_y, and the measurement results in a linear relationship between ω and B_y. The intercept is the residual magnetic field inside the magnetometer. (**a**) Probe light path A in Figure 2 (P_x transient response). (**b**) Probe light path B in Figure 2 (P_z transient response). An expanded version of the inset is at the right of the image.

The experimental data under different B_y in Tables 3 and 4 were compared. The polarization obtained using the P_z response was smaller than those obtained from the P_x response. This was because the probe beam measured the average of the polarization rates in the path it goes through. The polarization distribution was proportional to the light intensity distribution in the cell. The R_{op} in the cell was different. When the probe light was perpendicular to the pump light (path A), its path was only affected by the integral of the laser Gaussian light field. When the probe light was parallel to the pump light (path B), the probe path was affected by the integration of the light field formed by the large OD. In our experiment, because the OD had a greater impact on the path integral of polarization, the polarization measured by the P_z response decreased. At low frequencies, the phase-locked capability caused error bars to increase as B_y decreased. In addition, the depolarization caused by the window plate and the cell and the pumping effect of the probe light on the atoms resulted in measurement errors.

Table 3. P_x response measurement results under different magnetic fields.

Magnetic Fields B_y(nT)	Precession Frequency (Hz)	Slow-Down Factor $q(P)$	Polarization
1	3.33 ± 0.35	8.42 ± 0.81	0.57 ± 0.11
4	10.35 ± 0.33	10.82 ± 0.33	0.31 ± 0.04
7	17.14 ± 0.21	11.43 ± 0.14	0.24 ± 0.01
10	23.55 ± 0.20	11.89 ± 0.10	0.18 ± 0.01

Table 4. P_z response measurement results under different magnetic fields.

Magnetic Fields B_y(nT)	Precession Frequency (Hz)	Slow-Down Factor $q(P)$	Polarization
1	3.13 ± 0.41	8.94 ± 1.04	0.51 ± 0.13
4	10.07± 0.52	11.12 ± 0.44	0.27 ± 0.05
7	16.97 ± 0.22	11.55 ± 0.15	0.23 ± 0.02
10	23.23 ± 0.24	12.06 ± 0.12	0.16 ± 0.02

4.2. Polarization Spatial Distribution Measurement

Inhomogeneity in the polarization distribution within the SERF magnetometer is caused by the inhomogeneous light field distribution within the cell, as described in the previous section [37]. The high density of alkali vapor absorbs the circularly polarized light as it passes through the cell, causing an attenuation of the light intensity in the longitudinal plane (yz plane). According to the Lambert–Beer absorption law, the pumping rate is related to the distance in the z direction by the following equation [17]:

$$R_{op}(z) = R_{rel}W\left[\frac{R_p}{R_{rel}}\exp\left(-n\sigma(v)z + \frac{R_p}{R_{rel}}\right)\right], \tag{15}$$

where $\sigma(v)$ is the absorption cross-section as a function of the laser frequency, n is the density of the alkali atoms vapor, z is the pump light's propagated distance in the cell, and the initial pumping rate R_p is proportional to the pump light intensity I_{pump}.

For the transverse plane (xy plane), the optical field inhomogeneity was mainly caused by the laser's Gaussian light intensity distribution. The pumping rate was related to the distance in the x and y directions by the following equation:

$$R_{op}(x,y) = R_p e^{\left(\frac{-2(x^2+y^2)}{r^2}\right)}, \tag{16}$$

where r is the pump light beam radius. The atoms had sufficient collision mixing in the SERF state, and the initial electron spin polarization P_0 can be written as follows:

$$P_0 = \frac{R_{op}}{R_{op} + R_{\text{rel}}}. \tag{17}$$

When only the B_y magnetic field exists, the steady-state solution of Equation (6) is as follows:

$$P_z = \frac{P_0}{1 + \left(\frac{\gamma_e B_y}{R_{op}+R_{rel}}\right)^2}. \tag{18}$$

Figure 5 shows the spin polarization distribution in the z axis of the pump light propagation direction and the x and y axes of the pump light cross-section direction, which is fitted by Equations (16)–(18). Figure 5a is measured from the transient response of the magnetometer obtained using the experimental device optical A path, and Figure 5b,c are measured using the optical B path. The origin corresponds to the center of the cell. DMD is used for distribution measurement with a distance of 0.25 cm between each measuring point. The distribution measurement method can accurately locate the measurement position. In our measurement range of −0.75 to 0.75 cm, the polarization decreased by 98.18% due to the light intensity attenuation caused by alkali metal absorption and by 53.36% due to the Gaussian light intensity distribution. Although light intensity absorption is the primary cause of polarization inhomogeneity, Gaussian light causes transverse polarization inhomogeneity, which affects the sensitivity of magnetometer triaxial measurements and the application of biomagnetic detection.

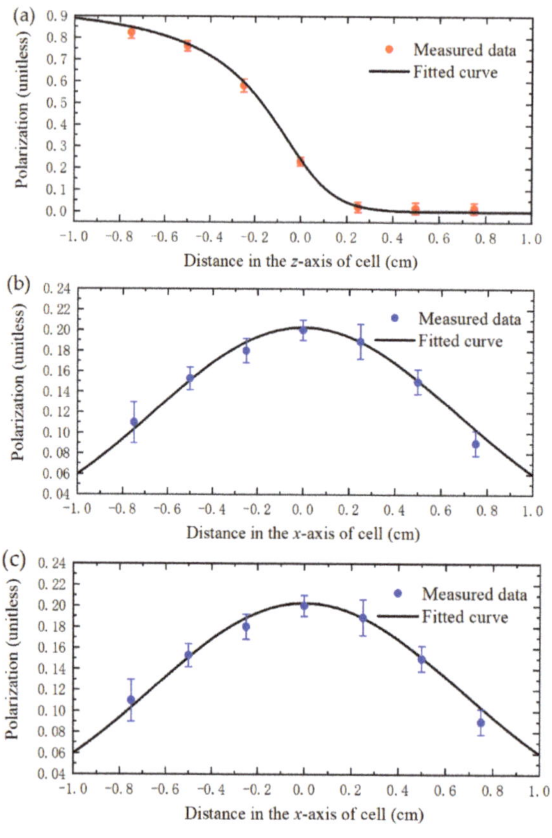

Figure 5. Experimental measurement result of the spin polarization distribution in the z axis in the pump light propagation direction and in the x and y axes in the pump light cross-section direction. The measuring spatial resolution is 0.25 cm, with a measuring range of −0.75–0.75 cm. The error bars represent the expanded uncertainties of the results. The solid line is the fitted curve obtained by Equations (16)–(18). (**a**–**c**) is the polarizability measured along the x-, y-, z- triaxial axis respectively.

Figure 6 shows the measurement results of the spatial polarizability distribution of xy and yz planes with a spatial resolution of 0.25 cm using the experimental process shown in Figure 4. We measured 7 × 7 space points on each side, and each space point was measured three times. Table 2 shows the experimental conditions. We use δ to represent the measured polarization inhomogeneity, which can be described as follows:

$$\delta = \sqrt{\sum\left(\frac{(P-\overline{P})}{\overline{P}}\right)^2/(m-1)}, \tag{19}$$

where m is the number of measurement points. Within our measurement range of $0.75\ \text{cm}^2 \times 0.75\ \text{cm}^2$, the nonuniformity of the xy plane is δ = 1.04, and that of the yz plane is δ = 1.82. The Gaussian light field also affects the nonuniformity of the yz plane. The measurement results of spatial distribution were affected by the interatomic diffusion effect, which could be further suppressed by increasing the pressure in the vapor cell. Based on the proposed method for measuring the spatial distribution of transverse and longitudinal polarization, the polarization inhomogeneity in the yz plane could be further suppressed by a hybrid optical pumping technique [38], and the polarization inhomogeneity in the xy plane can be further reduced by beam shaping technology [39] to further improve the sensitivity of the SERF magnetometer and the application performance of the SERF magnetometer.

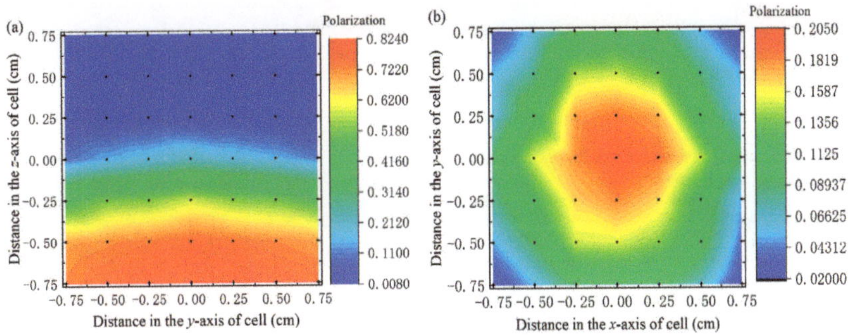

Figure 6. Experimental measurement result of spatial polarizability distribution of xy and yz planes with a spatial resolution of 0.25 cm. Each side measures 7 × 7 space points, and each space point is measured three times. (**a**) The measurement results of the spatial polarizability distribution of xy plane, (**b**) The measurement results of the spatial polarizability distribution of yz plane.

5. Conclusions

In this study, we proposed a novel method for measuring both the transverse and longitudinal plane spatial distributions of alkali metal polarization for a SERF magnetometer using the transient response of the magnetometer after modulating the pumped light using a chopper at 1.6 Hz. The dynamics equation and working principle were demonstrated. The main conclusions that could be drawn were:

- By using the P_x and P_z transient responses, the slowing factor $q(P)$ at the center point of the cell at different B_y magnetic fields was obtained, and the polarization was calculated based on the fast spin-exchange interaction effect.
- The device's residual magnetic field B_{res} was obtained from the linear relationship of precession frequency ω_{pr} and B_y.
- Measuring the one-dimensional polarization distribution of the x, y, and z axes with 0.1 cm resolution. The measurement results were consistent with the Lambert–Beer absorption law and the Gaussian distribution law.
- Two-dimensional spatial distribution measurements of the polarization of the xy and yz planes with a measurement range of $0.75\ \text{cm}^2 \times 0.75\ \text{cm}^2$. The nonuniformity of the xy plane was 1.04, and that of the yz plane was 1.82.

Based on the results of the proposed measurement method of spin polarization spatial distribution, further compensation can improve the application consistency of multi-channel magnetic field measurements and improve the sensitivity of single-channel differential measurements. Compared with existing methods for measuring polarization distribution, this method has the following advantages:

- It can measure three-axis spatial distribution spin polarization in the SERF region.
- It is independent of the OD and does not require a large background or RF field for applications suitable for both low and high polarization.
- It uses a square cell to avoid measurement errors caused by optical aberrations due to the circular cell's curvature.
- DMD distribution measurements provide high spatial resolution and are accurate in their positioning.

Our proposed measurement method applies equally to NMR magnetometers, all-optical atomic magnetometers, and comagnetometers.

Author Contributions: Conceptualization, X.F. and J.L.; methodology, K.W.; validation, X.F., Y.M., W.F. and J.L.; formal analysis, X.F.; investigation, X.F. and J.L.; data curation, X.F.; writing—original draft preparation, X.F.; writing—review and editing, Y.M. and W.F.; project administration, W.Q., Y.Z. and Z.X.; funding acquisition, W.Q. and W.F. All authors have read and agreed to the published version of the manuscript.

Funding: This work was supported by the National Science Fund for Distinguished Young Scholars, 61925301, and the National Natural Science Foundation of China under Grant No. 61903013, No. 62103026, and No. 61975005.

Institutional Review Board Statement: Not applicable.

Informed Consent Statement: Not applicable.

Data Availability Statement: The data presented in this study are available on reasonable request from the corresponding author.

Conflicts of Interest: The authors declare no conflict of interest.

References

1. Romalis, M.V. Hybrid optical pumping of optically dense alkali-metal vapor without quenching gas. *Phys. Rev. Lett.* **2010**, *105*, 243001. [CrossRef]
2. Lu, J.X.; Zhang, S.W.; Zhou, Y.; Yan, Y.G.; Lu, F.; Wang, K.; Zhai, Y.Y.; Ye, M. Optimal buffer gas pressure in dual-beam spin-exchange relaxation-free magnetometers. *Sens. Actuator A Phys.* **2022**, *347*, 113928. [CrossRef]
3. Lu, F.; Lu, J.X.; Li, B.; Yan, Y.G.; Zhang, S.W.; Yin, K.F.; Ye, M.; Han, B.C. Triaxial Vector Operation in Near-Zero Field of Atomic Magnetometer with Femtotesla Sensitivity. *IEEE Trans. Instrum. Meas.* **2022**, *71*, 1501210. [CrossRef]
4. Wei, K.; Zhao, T.; Fang, X.J.; Xu, Z.T.; Zhai, Y.Y.; Quan, W. Accurate Polarimetry of Hybrid K-Rb and ^{21}Ne Atoms Based on Spin-Exchange Interactions. *IEEE Sens. J.* **2021**, *21*, 5879. [CrossRef]
5. Xu, Z.T.; Wei, K.; Zhao, T.; Cao, Q.; Liu, Y.; Hu, D.; Zhai, Y.Y. Fast Dynamic Frequency Response-Based Multiparameter Measurement in Spin-Exchange Relaxation-Free Comagnetometers. *IEEE Trans. Instrum. Meas.* **2021**, *70*, 7007. [CrossRef]
6. Gill, P. Precision measurements: Optical clocks coming of age. *Nature* **2000**, *407*, 6804. [CrossRef]
7. Jiménez-Martínez, R.; Kennedy, D.J.; Rosenbluh, M.; Donley, E.A.; Knappe, S.; Seltzer, S.J.; Ring, H.L.; Bajaj, V.S.; Kitching, J. Optical hyperpolarization and NMR detection of ^{129}Xe on a microfluidic chip. *Nat. Commun.* **2014**, *5*, 3908. [CrossRef] [PubMed]
8. Yao, K.W.; Liu, M.; Zheng, Z.Y.; Shih, T.; Xie, J.Y.; Sun, H.J.; Chen, Z. Automatic Shimming Method Using Compensation of Magnetic Susceptibilities and Adaptive Simplex for Low-Field NMR. *IEEE Trans. Instrum. Meas.* **2021**, *70*, 6007012. [CrossRef]
9. Sherson, J.; Krauter, H.; Olsson, R.K.; Julsgaard, B.; Polzik, E.S. Quantum memory and teleportation using macroscopic gas samples. *Phys. B At. Mol. Opt. Phys.* **2008**, *41*, 223001. [CrossRef]
10. Vasilakis, G.; Shen, H.; Jensen, K.; Balabas, M.; Salart, D.; Chen, B.; Polzik, E.S. Generation of a squeezed state of an oscillator by stroboscopic back-action-evading measurement. *Nat. Phys.* **2015**, *11*, 389. [CrossRef]
11. Dang, H.B.; Maloof, A.C.; Romalis, M.V. Ultra-high sensitivity magnetic field and magnetization measurements with an atomic magnetometer. *Appl. Phys. Lett.* **2010**, *97*, 15. [CrossRef]
12. Sheng, J.; Wan, S.G.; Sun, Y.F.; Dou, R.S.; Guo, Y.H.; Wei, K.Q.; He, K.Y.; Qin, J.; Gao, J.H. Magnetoencephalography with a Cs-based highsensitivity compact atomic magnetometer. *Rev. Sci. Instrum.* **2017**, *88*, 9. [CrossRef] [PubMed]

13. Boto, E.; Holmes, N.; Leggett, J.; Roberts, G.; Shah, V.; Meyer, S.S.; Munoz, L.D.; Mullinger, K.J.; Tierney, T.M.; Bestmann, S.; et al. Moving magnetoencephalography towards real-world applications with a wearable system. *Nature* **2018**, *555*, 7698. [CrossRef] [PubMed]
14. Zhang, J.; Liu, K.N.; Zhang, J.F.; Wang, Z.J.; Shang, J.T. Magnetocardiography Measurements by Microfabricated Atomic Magnetometer with a 3-D Spherical Alkali Vapor Cell. *IEEE Trans. Instrum. Meas.* **2021**, *70*, 4007907. [CrossRef]
15. Li, Z.; Wakai, R.T.; Walker, T.G. Parametric modulation of an atomic magnetometer. *Appl. Phys. Lett.* **2006**, *89*, 13. [CrossRef]
16. Baranga, B.A.; Appelt, S.; Erickson, C.J.; Young, A.R.; Happer, W. Alkali-metal-atom polarization imaging in high-pressure optical-pumping cells. *Phys. Rev. A* **1998**, *58*, 3. [CrossRef]
17. Seltzer, S.J. Developments in Alkali-Metal Atomic Magnetometry. Ph.D. Thesis, Princeton University, Princeton, NJ, USA, 2008.
18. Gillen-Christandl, K.; Gillen, G.D.; Piotrowica, M.J.; Saffman, M. Comparison of Gaussian and super Gaussian laser beams for addressing atomic qubits. *Appl. Phys. B* **2016**, *122*, 131. [CrossRef]
19. Appelt, S.; Unlu, T.; Zilles, K.; Shah, N.J.; Baer, S.; Halling, H. Experimental studies of rubidium absolute polarization at high temperatures. *Appl. Phys. Lett.* **1999**, *75*, 3. [CrossRef]
20. Young, A.R.; Appelt, S.; Baranga, A.B.; Erickson, C.; Happer, W. Three-dimensional imaging of spin polarization of alkali-metal vapor in optical pumping cells. *Appl. Phys. Lett.* **1997**, *70*, 23. [CrossRef]
21. Lancor, B.; Babcock, E.; Wyllie, R.; Walker, T.G. Circular dichroism of RbHe and RbN_2 molecules. *Phys. Rev. A* **2010**, *82*, 43435. [CrossRef]
22. Gusarov, A.; Levron, D.; Baranga, A.B.A.; Paperno, E.; Shuker, R. An all-optical scalar and vector spin-exchange relaxation-free magnetometer employing on-off pump modulation. *J. Appl. Phys.* **2011**, *109*, 130801. [CrossRef]
23. Li, R.J.; Quan, W.; Fang, J.C. Polarization measurement of Cs using the pump laser beam. *IEEE Photon. J.* **2017**, *9*, 1. [CrossRef]
24. Fang, X.J.; Wei, K.; Zhao, T.; Zhai, Y.Y.; Ma, D.Y.; Xing, B.Z.; Xiao, Z.X. High spatial resolution multichannel optically pumped atomic magnetometer based on spatial light modulator. *Opt. Express* **2020**, *28*, 26447–26460. [CrossRef] [PubMed]
25. Zhao, J.P.; Ding, M.; Lu, J.X.; Yang, K.; Ma, D.Y.; Yao, H.; Liu, G. Determination of Spin Polarization in Spin-Exchange Relaxation-Free Atomic Magnetometer Using Transient Response. *IEEE Trans. Instrum. Meas.* **2022**, *69*, 845. [CrossRef]
26. Kim, K.; Begus, S.; Xia, H.; Lee, S.K.; Jazbinsek, V.; Trontelj, Z.; Romalis, M.V. Multi-channel atomic magnetometer for magnetoencephalography: A configuration study. *Neuroimage* **2014**, *89*, 143. [CrossRef]
27. Ito, Y.; Sato, D.; Kamada, K.; Kobayashi, T. Measurements of Magnetic Field Distributions with an Optically Pumped K-Rb Hybrid Atomic Magnetometer. *IEEE Trans. Magn.* **2014**, *50*, 1. [CrossRef]
28. Ressler, N.W.; Sands, R.H.; Stark, T.E. Measurement of Spin-Exchange Cross Sections for ^{133}Cs, ^{87}Rb, ^{85}Rb, ^{39}K, and ^{23}Na. *Phys. Rev.* **1969**, *84*, 102. [CrossRef]
29. Happer, W.; Tang, H. Spin-Exchange Shift and Narrowing of Magnetic Resonance Lines in Optically Pumped Alkali Vapors. *Phys. Rev. Lett.* **1973**, *31*, 273. [CrossRef]
30. Savukov, I.M.; Romalis, M.V. Effects of spin-exchange collisions in a high-density alkali-metal vapor in low magnetic fields. *Phys. Rev. A* **2005**, *71*, 23405. [CrossRef]
31. Happer, W.; Tam, A.C. Effect of rapid spin exchange on the magnetic-resonance spectrum of alkali vapors. *Phys. Rev. A* **1977**, *16*, 1877. [CrossRef]
32. Anderson, L.W.; Pipkin, F.M.; Baird, J.C., Jr. Hyperfine Structure of Hydrogen, Deuterium, and Tritium. *Phys. Rev.* **1960**, *121*, 1864. [CrossRef]
33. Huang, H.C.; Dong, H.F.; Hu, X.Y.; Chen, L.; Gao, Y. Three-axis atomic magnetometer based on spin precession modulation. *Appl. Phys. Lett.* **2015**, *107*, 182403. [CrossRef]
34. Duan, L.H.; Fang, J.C.; Li, R.J.; Jiang, L.W.; Ding, M.; Wang, W. Light intensity stabilization based on the second harmonic of the photoelastic modulator detection in the atomic magnetometer. *Opt. Express* **2015**, *23*, 32481. [CrossRef]
35. Ma, D.Y.; Lu, J.X.; Fang, X.J.; Yang, K.; Wang, K.; Zhang, N.; Han, B.C.; Ding, M. Parameter modeling analysis of a cylindrical ferrite magnetic shield to reduce magnetic noise. *IEEE Trans. Ind. Electron.* **2022**, *69*, 991. [CrossRef]
36. Lu, J.X.; Quan, W.; Ding, M.; Qi, L.; Fang, J.C. Suppression of light shift for high-density alkali-metal atomic magnetometer. *IEEE Sens. J.* **2019**, *19*, 492. [CrossRef]
37. Ledbetter, M.P.; Savukov, I.M.; Acosta, V.M.; Budker, D.; Romalis, M.V. Spin-exchange-relaxation-free magnetometry with Cs vapor. *Phys. Rev. A* **2008**, *77*, 33408. [CrossRef]
38. Ito, Y.; Ohnishi, H.; Kamada, K.; Kobayashi, T. Sensitivity improvement of spin-exchange relaxation free atomic magnetometers by hybrid optical pumping of potassium and rubidium. *IEEE Trans. Magn.* **2011**, *47*, 3550. [CrossRef]
39. Chen, X.; Fang, X.J.; Ma, D.; Liu, Y.; Cao, L.; Zhai, Y.Y. Optimization of beam shaping for ultrasensitive inertial measurement using a phase-only spatial light modulator. *Appl. Opt.* **2022**, *61*, C55. [CrossRef]

MDPI AG
Grosspeteranlage 5
4052 Basel
Switzerland
Tel.: +41 61 683 77 34

Photonics Editorial Office
E-mail: photonics@mdpi.com
www.mdpi.com/journal/photonics

www.ingramcontent.com/pod-product-compliance
Lightning Source LLC
Chambersburg PA
CBHW062100220926
51850CB00010B/2841
* 9 7 8 3 7 2 5 8 4 6 9 3 1 *